Infrastructure Management and Construction

Edited by Samad M.E. Sepasgozar, Faham Tahmasebinia and Sara Shirowzhan

Published in London, United Kingdom

IntechOpen

Supporting open minds since 2005

Infrastructure Management and Construction
http://dx.doi.org/10.5772/intechopen.78484
Edited by Samad M.E. Sepasgozar, Faham Tahmasebinia and Sara Shirowzhan

Contributors
Mohammadsoroush Tafazzoli, Vittorio Rosato, Sonia Giovinazzi, Antonio Di Pietro, Alberto Tofani, Maurizio Pollino, Gregorio D'Agostino, Luigi La Porta, Rex Asibuodu Ugulu, Stephen Allen, Andrew Arewa, Samad M.E. Sepasgozar, Michael Schorr, Arturo Gonzalez, Benjamin Valdez, Alejandro Mungaray, Martin Hromada, Kateřina Víchová

First published in London, United Kingdom, 2020 by IntechOpen
IntechOpen is the global imprint of INTECHOPEN LIMITED, registered in England and Wales, registration number: 11086078, 7th floor, 10 Lower Thames Street, London, EC3R 6AF, United Kingdom
Printed in Croatia

British Library Cataloguing-in-Publication Data
A catalogue record for this book is available from the British Library

Additional hard and PDF copies can be obtained from orders@intechopen.com

Infrastructure Management and Construction
Edited by Samad M.E. Sepasgozar, Faham Tahmasebinia and Sara Shirowzhan
p. cm.
Print ISBN 978-1-78984-548-8
Online ISBN 978-1-78984-549-5
eBook (PDF) ISBN 978-1-78985-316-2

For EU product safety concerns:
IN TECH d.o.o., Prolaz Marije Krucifikse Kozulić 3, 51000 Rijeka, Croatia, info@intechopen.com or visit our website at intechopen.com.

Meet the editors

Dr. Samad M.E. Sepasgozar is a senior lecturer in digital construction and property management who has published more than 100 trending papers in the last seven years. He has been granted funds as chief investigator for different projects on the application of advanced technologies. Dr. Sepasgozar has a strong background in developing smart technologies for the built environment. His work has been recognized by world leaders, as evidenced by "best paper'" awards for some of his publications in high-quality journals and world-class conferences. He is an editorial board member for a leading Q1 journal, and guest editor and reviewer of many high-ranked journals focusing on innovation and smart technologies.

Dr. Faham Tahmasebinia has a PhD in Civil Structural Engineering from the University of Sydney, Australia, and a PhD in Rock Mechanics from the University of New South Wales, Australia. He also holds a MEng in Structural Engineering from the University of Wollongong, Australia. Dr. Tahmasebinia has been teaching different structural and constructional management engineering subjects at the University of Sydney since 2009. His main research expertise is in finite elements, discrete elements, and analytical modeling in different conditions.

Dr. Sara Shirowzhan is a lecturer of City Analytics in the Faculty of Built Environment at the University of New South Wales (UNSW), Sydney, Australia. She completed her PhD in Geomatic Engineering at the School of Civil and Environmental Engineering, UNSW. In 2007, she started teaching in built environment disciplines. Currently, she teaches GIS, BIM, and major project courses at undergrad and postgrad levels. She has supervised twenty-two research students so far and has been involved in university- and government-funded research projects. Her areas of interest include advanced technologies for sustainable and smart built environments including laser scanning technologies, nD BIM, nD GIS, big data analysis, VR, and AR.

Contents

Preface

Infrastructure construction and maintenance are the key requirements of regional and national development in different countries. At the same time, infrastructure projects require greater investment compared to smaller building construction projects. Therefore there is a need to focus on this topic and conduct more investigations in the field of infrastructure projects. This book presents some key concepts and practices around the theme of critical infrastructure.

Chapter 1 describes the key concepts of infrastructure management, construction, structure, and Industry 4.0, which includes mapping and monitoring technologies. This chapter briefly presents a set of selected infrastructure projects in a metropolitan area. It also presents examples of technologies relevant to Industry 4.0, such as temporal airborne light detection and ranging (lidar), which can be used for analyzing 3D urban changes.

Chapter 2 discusses four categories of sustainability maintenance of critical infrastructures: (1) minimizing adverse impacts of the infrastructure on people through maintenance, (2) keeping the maintenance operations sustainable, (3) allocating sustainable material throughout the maintenance process, and (4) protecting and restoring the environment in maintenance operations. In each category, some of the best practices and methods are discussed. Also highlighted is the significance of considering the hosting community's concerns, the managers' determination to maintain sustainable performance, sustainable handling of materials, sustainable maintenance operations, and continuous restoration of damages caused by the infrastructure were highlighted.

Chapter 3 presents and proposes the use of a complex simulation model, called the reconfiguration simulator (RecSIM), enabling the evaluation of the effectiveness of resilience enhancement strategies for electric distribution networks and the required resources to implement them. The focus is on one specific attribute of resilience, which refers to the promptness and efficiency which recover the service functionality after a crisis event by managing and deploying the available resources rapidly and effectively. The chapter discusses that the RecSIM allows estimating how and to what extent technological, topological, and management issues might improve electrical distribution networks' functionality after the occurrence of accidental faults, accounting for interdependency issues and reconfiguration possibilities.

Chapter 4 presents the concept and meaning of construction productivity and the techniques used in measuring workforce productivity in the construction industry. It discusses three major methods for monitoring and measuring this productivity. The first relates to visual recording that requires taught watchers to be available on location to monitor and record work by specialists. The second is the physical recording technique, which involves direct surveillance that includes a qualified observer observing the site for the full duration of the work day using a work study method or work measurement. The third technique discussed is the use of questionnaires and interview surveys, which involve information gathering through an interview with supervisors and workers working in the construction industry.

Chapter 5 refers to one of the significant sectors of critical infrastructure, which is undoubtedly healthcare. It is vital for health services to be able to perform their function, even in times of crisis. The crisis of naturogenic character has the so-called cascade effect, which causes other extraordinary events and crises. An example of this may be a windstorm resulting in a power outage. A significant impact of the power supply outage in healthcare is observed in hospitals. There is currently no assessment tool to set the hospital's readiness for a power outage. The aim of the chapter is to analyze the current state of the crisis preparedness of the hospital.

Chapter 6 presents a brief review of ancient and modern bridges, including the process of design, material selection, construction, and maintenance. It discusses that each structure has different requirements to cover, such as span clearage, traffic flow, geometry, and characteristics of the place to build; therefore, a great variety of bridges can be developed. Common materials used in construction are structural steel, reinforced concrete, pre-stressed concrete, or post-tensioned concrete. Depending on the structural behavior of each type of bridge, there will be a maximum clear span to cover, which depends directly on the project's budget. Prevention and control of degradation processes is achieved by effective maintenance methods applying protection technology such as paints, coatings, and cathodic protection.

Dr. Samad M.E. Sepasgozar and Dr. Sara Shirowzhan
The University of New South Wales,
Sydney, Australia

Dr. Faham Tahmasebinia
The University of Sydney,
Australia

Chapter 1

Introductory Chapter: Infrastructure Management, Construction, Structure and Industry 4.0

Samad M.E. Sepasgozar, Sara Shirowzhan and Faham Tahmasebinia

1. Introduction

Infrastructure has large capacity to enhance the efficiency of cities and the productivity of the national economy. It has a wide range of benefits from different perspectives including social, economy and environment in both short and long terms. The main objectives of an infrastructure are to improve national, interregional and international network and connectivity for people, business and trades. Infrastructure development is one of the most important functions of governments all around the world since it has a great impact of the national economy.

This book presents some concepts and practices around the theme of critical infrastructure. The following sections introduce (i) the concept of critical infrastructure management, (ii) structural design for concrete elements of infrastructure, and (iii) Industry 4.0 and digital data, which may be used for infrastructure management including mapping and monitoring technologies.

2. Definitions and concepts

Many countries are planning to invest on infrastructure projects. For example, the Australian government planned to invest over $75 billion for developing transportation network including tunnelling, roadways and railworks across the country in a period of 10 years.

Infrastructure generally refers to any foundation, system or basic physical for social development. It can be a basic structure, system or service such as highways, streets, roads, bridges, mass transit infrastructures (e.g. airports), dam, reservoirs, water supply and resources, waste and waste water infrastructure, hydroelectric plants, massive irrigation systems, telecommunications facilities, power generation and transmission and hazardous waste removal and storages. This type of infrastructure is synonymous to economic infrastructure. Social infrastructure includes cultural, educational and healthcare facilities.

The ninth goal of the Sustainable Development Goals introduced by the United Nations Development Programme refers to 'industry, innovation and infrastructure'. Innovation refers to the use of any new creative idea, technology, system, process, practice, material artefact or a nontrivial change and improvements that bring about valuable and meaningful change with benefit to stakeholders [1, 2].

IntechOpen

Term and concept	Purpose and definition
Sustainable infrastructure	Gives access to energy, improves connectivity and mobility, increases the community benefits and balances social, economic and environmental needs. It also includes rehabilitation and reuse of current infrastructure
Smart city [3]	A modern high-tech city that uses intelligent and communication technologies to connect citizens, information and governance and manage resources, economy and city elements. Its purpose is to provide a sustainable, green, efficient and innovative environment and to increase mobility and the human life quality
Smart infrastructure (structural health monitoring)	Utilisation of new information and communication systems including connected sensors (Internet of Things) to monitor and manage infrastructure in a real-time manner. The sensors help to minimise disruption, costs and down time and maximise capacity, operational efficiency, effectiveness and sustainability
Critical infrastructure	Refers to assets, facilities, systems and processes required for a society to function. This may include heating facilities (e.g. natural gas, fuel), food production and distribution networks, water supply (water, sewage), hospitals, transportation systems (railway network, airports, harbours, fuel supply), electricity generation and distribution systems, telecommunication networks Critical infrastructure risks: risk = *threat* × *vulnerability* × *consequence*
Critical urban infrastructure	Refers to facilities and network systems at the urban scale such as water, sewerage, power lines, gas and telecommunications
Critical infrastructure sectors	A total of 16 sectors identified by Presidential Policy Directive 21 (PPD-21) are [4] transportation systems, water and wastewater systems, information technology, materials and waste, government facilities, healthcare and public health, financial services, food and agriculture, energy, emergency services, dams, defence industrial base, communication, critical manufacturing, commercial facilities and chemical sectors
Critical infrastructure resilience (CIR)	Refers to the ability of the city or country to anticipate, prevent and protect with a coordinated plan for the network, responsive and timely recovery actions, while circumvent threats provide minimum level of services

Table 1.
Definitions of basic terms and concepts.

It also mentions that economic growth highly depends on the investment on infrastructure and innovation. There are several terms and concepts related to infrastructure, which will be introduced in **Table 1**.

Australian investment plan is an example of infrastructure investment plans, which are committed in developing transport infrastructures over a decade. The investment plan presents a positive reform and roadmap for a period of 15 years for this country. The main objectives of the investment plan are to improve

Figure 1.
Woolgoolga to Ballina in October 2018 (project ID: INF12): (a) bridge over Clarence River and (b) bridge over Richmond River. (courtesy: www.pacifichighway.nsw.gov.au).

Project ID	Infrastructure project	Short description
INF01	WestConnex (including different stages)	33-km-long underground motorway including 19 km of new tunnels, widening 7.5 km of the existing M4, estimated project cost $10–$45 billion including land acquisitions and network extensions
INF02	M4 Western	46-km-long dual carriageway motorway
INF03	NorthConnex	9-km-long dual road tunnels, linking the M1 and M2 Motorways, estimated project cost $3 billion
INF04	The Northern Road upgrade	35-km-long, upgrades four to eight lanes in the Northern Road, estimated project cost $1.6 billion
INF05	M12 Motorway	A new east-west motorway in Sydney, major access route to the Western Sydney Airport, estimated project cost $1.2 billion (duration, 2020–2025)
INF06	Bringelly Road	10 km upgrading Bringelly Road from two lanes to a six-lane divided road, estimated project cost $509 million (duration, 2015–2020)
INF07	CityLink-Tullamarine widening project	Upgrading of the Tullamarine Freeway between the Melbourne Airport and Melrose Drive in Victoria state, estimated project cost $250 million (duration, 2016–2018)
INF08	Great Western Highway	Install concrete median barrier, install raised islands, safety upgrades at different locations of the Great Western Highway (NSW)
INF09	Bringelly Road upgrade	10-km-long, upgrades Bringelly Road from two lanes to a six-lane divided road, located near to Eastwood Road (NSW), $509 million (duration, 2015–2020)
INF10	Sydney Light Rail (including several projects or phases)	(i) CBD and South East Light Rail (12 km route including 19 stops) (ii) Inner West Light Rail (12.7 km route including 23 light rail stops, capacity to transport more than 9.7 million customers yearly) (iii) Newcastle Light Rail (iv) Light rail in Randwick
INF11	Badgerys Creek Airport	Sydney's second airport, estimated project cost $2.5 billion
INF12	Woolgoolga to Ballina (see **Figure 1**)	155-km-long, link in the Pacific Highway including 9 interchanges, 170 bridges and 350 connectivity structures
INF13	The new Northern Beaches Hospital	With 488 beds, 50 emergency bays, 14 operating theatres, 20 intensive care beds, 40 maternity beds

Note: Descriptions of the benefit of each project and any other specific project details can be obtained from websites of the Department of Infrastructure, Regional Development, New South Wales Government, New South Wales Roads and Maritime Services (RMS), and/or relevant city council, designers, consultants or contractors.

Table 2.
Selected infrastructure projects in a metropolitan.

connectivity of peopled and regions, interregional logistics, health, safety, security and effectivity to the national transport system. **Table 2** shows a list of examples of infrastructure projects to address these objectives.

3. Mechanism of prestressing

The structure of infrastructure can be a mix of different materials and elements [5]. This section focuses on one of the popular innovations which can be used in many different infrastructures. One of the most relevant concepts to structure is prestressing. Prestressing refers to the intentional creation of permanent stresses in an element of a structure, for the purpose of improving its strength and behaviour under various services and conditions [6]. Prestressed concrete is an engineering

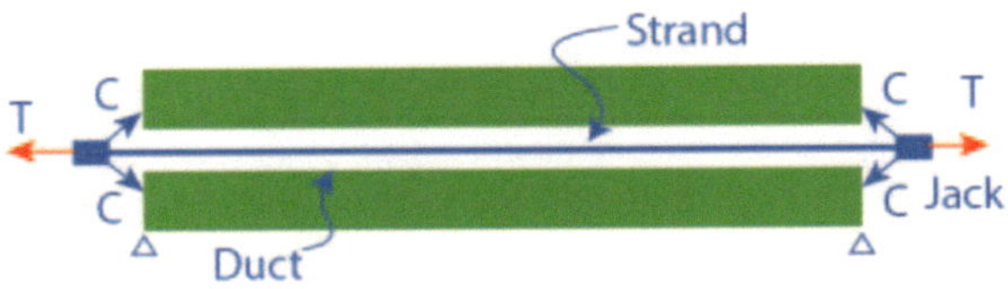

Figure 2.
A simplified illustration of the mechanism of prestressing.

innovation that improves many of the service and strength performance behaviours of reinforced concrete. This is a practical solution to the design of many engineering structures in a cost-effective manner. Prestressed concrete is a fastidious form of reinforced concrete. Prestressing engages the application of an initial compressive load to the structure to decrease internal tensile forces which may lead to controlling crack. The initial compressive load is imposed and sustained by highly tensioned steel reinforcement (tendons) reacting on the concrete. With eliminating cracking, a prestressed concrete section is significantly stiffer than the equal reinforced concrete section.

The tendons are cast within the concrete at first freely within ducts, which are grouted at a later stage to bond the tendons. Prior to grouting, the tendons are jacked to very high stresses. The jacking reactions are pressed against the ends of the concrete member and then transferred permanently to cast in end anchors. This is shown in a simplified form in **Figure 2**.

In practice the duct 'profile' will vary to suit the purpose of the member. The tendons are permanently 'prestressed' in tension; the concrete is permanently 'prestressed' in compression. High-tensile wire strands have $f_y \approx 1870$ MPa. Higher-strength concretes have $f_c' \approx 30$–50 MPa. Although strong in compression, concrete is weak in tension. **Table 3** presents the basic concepts in prestressed concrete structures.

The poor performance of the concrete in tension gave rise to the concept of 'prestressing' the concrete in order to overcome bending tensions and cracking produced under load, thereby retaining elastic behaviour. Early attempts at prestressing failed because of the shrinkage and creep strains in the concrete. For example, early reinforcing bars had yield stresses ≈ 200 MPa and could be pretensioned to a stress of order 120–150 MPa. The subsequent strains are of order

$$\varepsilon = \sigma/E = 120/200{,}000 = 600 \times 10^{-6} \tag{1}$$

where ε is the strain, σ is the stress and E is the module of elasticity.

The development of prestressed concrete is credited to Eugene Freyssinet of France [7], who in 1928 started using high-strength steel wires for prestressing. Such wires, with an ultimate strength as high as 1725 MPa and a yield point over 1240 MPa, are prestressed to about 1000 MPa. At this level of stress, the losses are a much smaller percentage of the prestress. For example, assuming a total creep and shrinkage strain of order 800×10^{-6}, this would result in a loss of prestress of order:

$$\sigma = (800 \times 10-6) \times (2 \times 105) = 160 \text{ MPa} \tag{2}$$

Hence, the residual prestress =1000–160 = 840 MPa or 84% of the original level. Practical development of prestressed concrete was made possible by further development of high-strength concrete and high-tensile wire strands through the twentieth century [8–11]. It was also necessary to develop reliable and economical

Element	Description
Strand	An element in which a number of high-tensile wires are woven together as a combined unit
Tendon	Generally defined as the wire, strand or bar (or any discrete group of wires, strands or bars) that is intended to be prestressed
Cable	Groups of tendons, collected together in a duct or anchorage
Prestressed	The prior stressing of both the concrete and the tendons prior to the service use of the element
Pretensioned	The tendons are tensioned (in a casting bed) prior to pouring of the concrete. The tendons are cut and bonded to the concrete after it attains strength. Used in precast construction
Post-tensioned	The tendons are laid in metal ducts in the concrete and tensioned after the concrete attains strength. This is the most commonly used system in buildings and other structures in Australia
Bonded tendons	Where the ducts which contain the tendons are filled with cement grout after the tendons are stressed—this effectively bonds them to the concrete like reinforcing steel. Tendons can subsequently develop further stresses under bending actions due to strain compatibility, enabling the full strength of the bonded tendons to be realised at ultimate strength
Unbonded tendons	Where the tendons are not grouted but are greased for corrosion protection and contained within a plastic sheath. This method is not permitted under as 3600 except for slabs on the ground. Unbonded tendons can be used in specialised applications such as cable-stayed bridge cables, for example, allowing the monitoring and replacement of individual strands
Immediate losses	(a) Short-term elastic strain of the concrete (progressive stressing) (b) Anchor wedge draw-in (c) Friction loss along the length of the ducts
Time-dependant losses	(a) Long-term creep and shrinkage strain of the concrete (b) Relaxation in the (seven wires twisted) strands

Table 3.
The basic concepts in the prestressed concrete structures including main elements of the structure.

methods of tensioning the strands and for the permanent end anchoring the strands to the concrete. Freyssinet [7] developed conical wedges for end anchorages and designed double acting jacks which tensioned the wires and then thrust the wedges into the anchor cones which held them. These early advances in prestressing were mainly European. Steel became the primary construction material in the USA, with the secondary use of structural concrete in buildings. However, the American engineer Lin [6] became a major contributor developing the load balancing theory during the 1950s. Australia, with a more developed concrete industry and low seismic activity, became a major advocate of prestressed concrete buildings (**Figure 3**).

Loss of prestress in the tendon force up to 40% or more of the jacking force can be experienced, depending upon the cable length, drapes, concrete shrinkage and so forth. This becomes important in relation to AS3600 [12]. It is necessary that a minimum amount of prestress remains in the strands, so that they may develop their full-strength capacity in strain compatibility with the concrete as it bends.

In addition to the longitudinal force P (or C) exerted on a prestressed member at the anchorages, transverse forces are also exerted on the concrete member wherever change of angle or curvature exists in the tendons. Other bending actions are also applied to the member where the member section changes due to

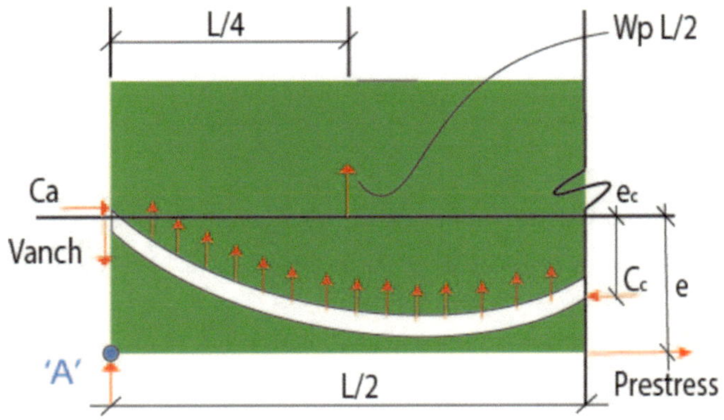

Figure 3.
A prestressed section for analysis.

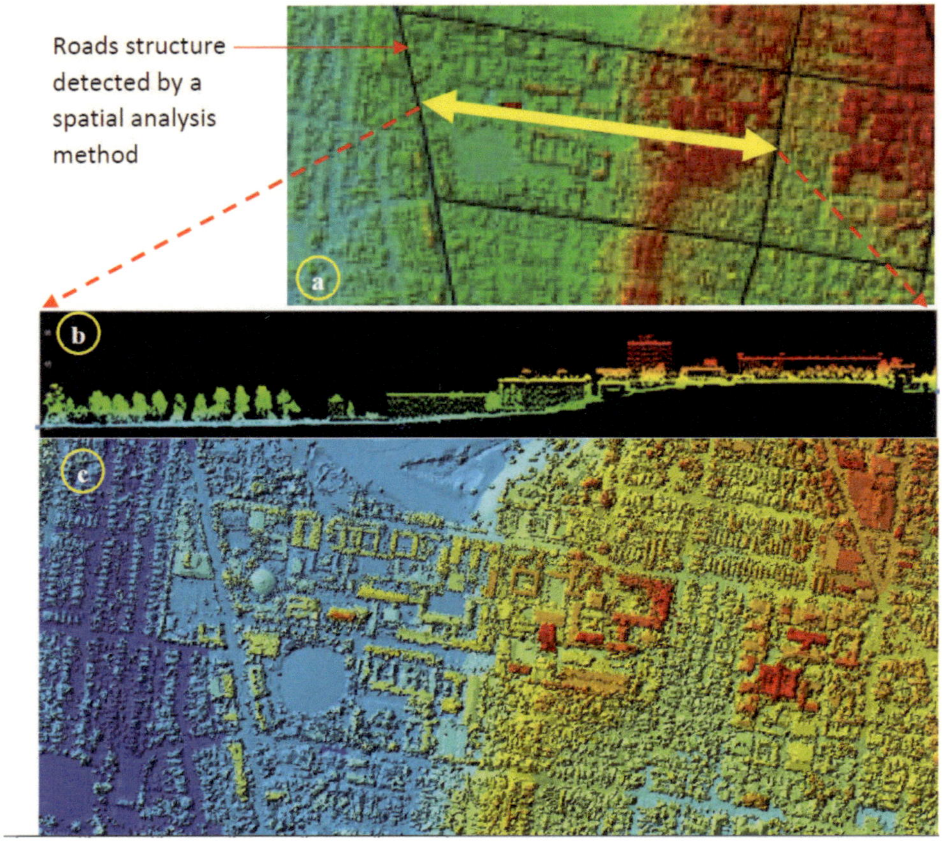

Figure 4.
Producing profile from an urban area to present building objects: (a) lidar data including road structure; (b) profile; (c) a selected lidar point cloud of Sydney areas produced by an UAV.

steps and so forth or P is applied eccentrically. For this analysis it is very helpful to remember that the concrete is a free body distinct from the tendon until the ducts are grouted, at which point they start to act together. This is then considered how the tendon imposes forces on the concrete free body at the anchorages and through its changes of direction. Consider the concrete in half of the beam as a free body, which has forces imposed on it from the tendon and the other half of the beam (**Figure 3**).

The forces on the concrete are:

- C_a and C_c in horizontal equilibrium
- $w_p L/2$ and V_{anch} in vertical equilibrium

Now determine e_c the eccentricity between the two C forces. Note that the P force in the tendon reacts against the concrete as C_a and there is no support reaction since there are no externally applied loads.

For moment equilibrium about **A**, and ignoring friction losses

$$C_c\ (= C_a = P) \times e_c\ = w_p L/2 \times L/4 = w_p\, L^2/8\ (= P \times e) \tag{3}$$

Now consider the elastic stresses in the concrete:

$$P/A\ (\text{pre - compression}) \pm M/Z\ (\text{uplift moment due to the cable}) \tag{4}$$

$$= P/A \pm P \times e \times y/I \tag{5}$$

In order to designing prestressed concrete structures, a proper understanding of structural performance at all stages of loading is important. This can be essential where it is a comprehensive knowledge of the design criteria specified in the relevant design standard, such as the minimum requirements to fulfil both ultimate and serviceability requirements.

4. Industry 4.0 technologies and digital data

This section presents different technologies and techniques which can be used for monitoring the structure of urban areas and infrastructures. The new trend of automation and data transferring process is called Industry 4.0. The core component of Industry 4.0 is digital data. Industry 4.0 refers to different technology applications using Internet of Things (IoT), cyber-physical systems, artificial intelligence, machine learning, cloud computing, machine-to-machine and human-to-machine communication and real-time technologies. The concept of Industry 4.0 is adopted by the construction industry. The concept of Construction Industry 4.0 can give better connectivity among construction supply chain stakeholders and real-time access to construction operation, enhancing safety, productivity and the quality of construction. Utilisation of digital technologies such as 3D printing [13], airborne lidar [3, 14, 15], hand-held laser scanners [16, 17], building information modelling (BIM) [18], wireless sensors [19] and automation is changing the infrastructure management including construction and maintenance.

For example, **Figure 4** shows the temporal airborne lidar which can be used for analysing 3D urban changes. Airborne lidar provides valuable digital data which is useful for analysis of the spatial pattern of building height and the proximity analysis with the road's hierarchy.

Spatial data mining (SDM) methods have been used to investigate the unknown relationships between spatial and nonspatial attributes of data sets [20] that may not be apparent using more basic data analysis techniques. The need for knowledge discovery and spatial data mining 'to extract unknown and unexpected information from spatial data sets' was suggested by Mennis and Guo [21]. Two popular SDM methods being used in geographic information systems (GIS) and remote sensing are spatial autocorrelation statistics [14, 21, 22] and nonparametric density estimation [23]. However, the potential of these SDM methods to explore urban height patterns using airborne lidar data has yet to be actively investigated. While a spatial autocorrelation statistic known as local Moran's I (LMI) is used to find the distribution pattern of building heights, the elevations of buildings were aggregated into large-sized cells using the mean elevation value of the included buildings [3, 22, 24].

Detection of the distribution pattern of clusters of relatively higher (CRH) buildings in a city with varying or heterogeneous heights is crucial for the spatial and temporal change analysis of vertical developments and for trend analysis of vertical urban compactness over time [15]. Detection of the CRH buildings in a city of heterogeneous heights is also essential for thermal urban modelling and urban heat island analyses because the level of heat produced by higher buildings is different from that by lower buildings [25].

Geostatistics is a useful technique for spatial analysis. It refers to statistics used to analyse spatial data and spatiotemporal data sets. Shirowzhan and Lim [22] utilised a spatial analysis procedure using temporal point clouds in advanced GIS. In this analysis a novel method examined ground elevation extraction in slant areas and building classifications. A relevant technique for measuring compactness in three-dimensional environment is Moran's *I* (MI) and *G* indices. Moran's I and *G* are global autocorrelation statistics, which computes the correlation between pairs of data points [14]. Autocorrelation can be calculated for a variable that changes over time, for linear spatial series and for two-dimensional spatial series. MI is an extended version of Pearson's product-moment correlation coefficient for a single variable [14]. Pearson's correlation between two variables x and y from n observations is defined as

$$\rho = \frac{\sum_{i=1}^{n} (x_i - \bar{x})(y_i - \bar{y})}{\left[\sum_{i=1}^{n} (x_i - \bar{x})^2 \sum_{i=1}^{n} (y_i - \bar{y})^2\right]^{1/2}} \tag{6}$$

where x and y are the mean values of x and y, respectively. For a univariate series, say x, MI will estimate the correlation between x_i and x_j.

For infrastructure monitoring, autocorrelation statistics can be applied to the variable describing the elevation of airborne lidar points in order to determine if x_i and x_j belong to the same class.

DOI: http://dx.doi.org/10.5772/intechopen.86142

Author details

Samad M.E. Sepasgozar[1]*, Sara Shirowzhan[1] and Faham Tahmasebinia[2]

1 Faculty of Built Environment, The University of New South Wales, Sydney, NSW, Australia

2 School of Minerals and Energy Resources Engineering, The University of New South Wales, Sydney, NSW, Australia

*Address all correspondence to: samad.sepasgozar@gmail.com

References

[1] Sepasgozar SME, Loosemore M, Davis SR. Conceptualising information and equipment technology adoption in construction. A critical review of existing research. Engineering Construction and Architectural Management. 2016;**23**(2):158-176. DOI: 10.1108/ECAM-05-2015-0083

[2] Sepasgozar SM, Davis SR, Li H, Luo X. Modeling the implementation process for new construction technologies: Thematic analysis based on Australian and US practices. Journal of Management in Engineering. 2018;**34**(3):05018005. DOI: 10.1061/(ASCE)ME.1943-5479.0000608

[3] Shirowzhan S, Sepasgozar SME. Spatial analysis using temporal point clouds in advanced GIS: Methods for ground elevation extraction in slant areas and building classifications. ISPRS International Journal of Geo-Information. 2019;**8**(3):120

[4] Directive PP. Critical infrastructure security and resilience. 2013. 21 (PPD-21). https://www.dhs.gov/sites/default/files/publications/ISC-PPD-21-Implementation-White-Paper-2015-508.pdf

[5] Tahmasebinia F, Sepasgozar SME, Blum H, Reddy KR, Alonso-Marroquin F, Gao Q, et al. The feasibility of constructing super-long-span bridges with new materials in 2050. In: Yaghoubi H, editor. Bridge Engineering. London: IntechOpen; 2018. DOI: 10.5772/intechopen.75298. ISSN: 978-1-78923-105-2

[6] Lin T. Strength of continuous prestressed concrete beams under static and repeated loads. Journal Proceedings. 1955;**51**:1037-1059

[7] Freyssinet E. Lecture on pre-stressed concrete: Principles & applications. Joint meeting with societe des ingenieurs civils de France (British section). 17 november 1949. Journal of the Institution of Civil Engineers. 1950;**33**(4):331-380

[8] Middleton TE. Support for elongated reinforcing members in concrete structures. Google Patents; 1972

[9] Vernon WJ. Method for pre-stressing concrete. Google Patents; 1963

[10] Cooke N, Park R, Yong P. Flexural strength of prestressed concrete members with unbonded tendons. PCI Journal. 1981;**26**(6):52-80

[11] Orbovic N, Elgohary M, Lee N, Blahoianu A. Tests on reinforced concrete slabs with pre-stressing and with transverse reinforcement under impact loading (5-2015). In: 20th International Conference on Structural Mechanics in Reactor Technology. 2009. p. 100

[12] Gilbert RI, Mickleborough NC, Ranzi G. Design of Prestressed Concrete to AS3600-2009. Boca Raton, Florida, USA: CRC Press; 2016. ISSN: 1466572779

[13] Tahmasebinia F, Niemelä M, Ebrahimzadeh Sepasgozar S, Lai T, Su W, Reddy K, et al. Three-dimensional printing using recycled high-density polyethylene: Technological challenges and future directions for construction. Buildings. 2018;**8**(11):165

[14] Shirowzhan S, Lim S, Trinder J. Enhanced autocorrelation-based algorithms for filtering airborne lidar data over urban areas. Journal of Surveying Engineering. 2015;**142**(2):04015008

[15] Shirowzhan S, Sepasgozar SME, Li H, Trinder J. Spatial compactness metrics and constrained voxel automata development for analyzing

3D densification and applying to point clouds: A synthetic review. Automation in Construction. 2018;**96**:236-249. DOI: 10.1016/j.autcon.2018.09.018

[16] Sepasgozar SM, Forsythe P, Shirowzhan S. Evaluation of terrestrial and mobile scanner technologies for part-built information modeling. Journal of Construction Engineering and Management. 2018;**144**(12):04018110. DOI: 10.1061/(ASCE)CO.1943-7862.0001574

[17] Sepasgozar S, Lim S, Shirowzhan S, Kim Y, Nadoushani ZM. Utilisation of a new terrestrial scanner for reconstruction of as-built models: A comparative study. In: Proceedings of the International Symposium on Automation and Robotics in Construction, ISARC; Vol. 32 2015; Vilnius Gediminas Technical University: Department of Construction Economics and Property. p. 1. DOI: 10.22260/ISARC2015/0027

[18] Hong Y, Hammad AW, Sepasgozar S, Akbarnezhad A. BIM adoption model for small and medium construction organisations in Australia. Engineering Construction and Architectural Management. 2019;**26**(2):154-183. https://doi.org/10.1108/ECAM-04-2017-0064

[19] Shirowzhan S, Sepasgozar SME, Zaini I, Wang C. An integrated GIS and Wi-fi based locating system for improving construction labor. Communications. Waterloo: IAARC Publications; 2017;**34**:1-8

[20] N. Chelghoum, K. Zeitouni, A. Boulmakoul, A decision tree for multi-layered spatial data, Advances in Spatial Data Handling, Springer, 2002, pp. 1-10. DOI: 10.1007/978-3-642-56094-1_1. ISSN: 978-3-642-62859-7

[21] Mennis J, Guo D. Spatial data mining and geographic knowledge discovery: An introduction. Computers, Environment and Urban Systems. 2009;**33**(6):403-408. DOI: 10.1016/j.compenvurbsys.2009.11.001

[22] Shirowzhan S, Lim S. Autocorrelation statistics-based algorithms for automatic ground and non-ground classification of Lidar data. In: ISARC Proceedings of the International Symposium on Automation and Robotics in Construction; vol. 31, Vilnius Gediminas Technical University: Department of Construction Economics. 2014. p. 1

[23] Srivastava AN, Oza NC, Stroeve J. Virtual sensors: Using data mining techniques to efficiently estimate remote sensing spectra. IEEE Transactions on Geoscience and Remote Sensing. 2005;**43**(3):590-600. DOI: 10.1109/TGRS.2004.842406

[24] Shirowzhan S, Trinder J. Building classification from lidar data for spatio-temporal assessment of 3D urban developments. Procedia Engineering. 2017;**180**:1453-1461. DOI: 10.1016/j.proeng.2017.04.308

[25] Giridharan R, Lau S, Ganesan S, Givoni B. Lowering the outdoor temperature in high-rise high-density residential developments of coastal Hong Kong: The vegetation influence. Building and Environment. 2008;**43**(10):1583-1595. DOI: 10.1016/j.buildenv.2007.10.003

Chapter 2

Maintaining the Sustainability of Critical Infrastructure

Mohammadsoroush Tafazzoli

Abstract

Critical infrastructures operations significantly affect the triple bottom lines of sustainability. Considering the dynamic nature of infrastructure and its surrounding environment, the interaction between them can dramatically change over time. This results in a deviation from the predictions used in the design and construction of the infrastructure. Therefore, the negative impacts of critical infrastructures on the environment, society, and economy can exacerbate throughout their service life. It is crucial to maintain these impacts within the desired limits. The measures that attempt to perpetuate a facility's adverse effects on the triple bottom lines of sustainability can be called sustainability maintenance. Regular maintenance operations of infrastructure create an opportunity to integrate sustainability maintenance into preventive, corrective, and periodic maintenances. This chapter discusses four categories of sustainability maintenance of critical infrastructures: (1) minimizing adverse impacts of the infrastructure on people through maintenance, (2) keeping the maintenance operations sustainable, (3) sustainable material allocation throughout the maintenance process, and (4) environmental protection and restoration in maintenance operations. In each category, some of the best practices and methods are discussed.

Keywords: critical infrastructure, maintenance, sustainability, environment, society, economy

1. Introduction

Critical infrastructures around the globe serve human beings in multiple aspects by supplying their essential needs for energy, water, food, healthcare, and transportation to name a few. These facilities are vital for the health, well-being, and economic growth of nations and are pivotal contributors to sustainable development. On the other hand, critical infrastructures can directly or indirectly affect the environment, economy, and society in a negative way [1–3]. For example, in the United States, almost 30% of all electricity is generated from coal. Coal power plants generate 42% of mercury emissions. This emission can damage the digestive, nervous, and immune systems, and is a critical threat to child development. While only 1/70th of a teaspoon of mercury deposited on a 25-acre lake can make the fish unsafe to eat [4]. As seen in **Table 1**, the EPA reports show that only in 2014, US coal plants generated 45,676 pounds of mercury [5].

The negative impacts of critical infrastructure are not only environmental. The ripple effect of ecological degradation can significantly affect the health and

Pollutant	Hg	SO2	NOx	PM*	Pb	Cd	CO	VOC**	As
Amount	45,676	3.1×10^6	1.5×10^6	197,286	41.2	9332	576,185	22,124	77,108
Unit	pounds	tons	tons	tons	tons	pounds	tons	tons	pounds

**Particulate matter.*
***Volatile organic compounds.*

Table 1.
Pollutants created by coal plants in the in 2014 in the USA

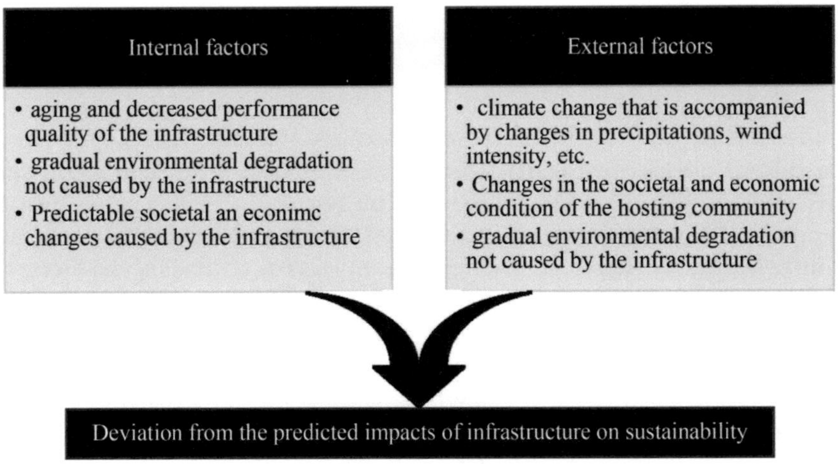

Figure 1.
Factors causing a deviation from the predicted interaction of infrastructure with its surrounding environment.

Power plant name	Location	Operation start	Shutdown year	Reason of shutdown
Crystal river	Crystal river, Florida	March 13, 1977	February 5, 2013	High price of repairs ($3.44 billion)
Kewaunee nuclear station	Kewaunee Wisconsin	December 1973	May 7, 2013	No buyer for the plant
San Onofre Nuclear Generating Station	San Clemente, California	1968	June 2013	Long time-repair and high restart costs
Clinton Nuclear Plant	Clinton, Illinois	1987	June 1, 2017	Becoming extremely uneconomical

Table 2.
The reasons for the shutdown of some nuclear power plants in the USA [8].

well-being of human beings as well as the economy of the hosting community. For instance, in the example above the pollutants generated by the coal-fired power plants contribute to cancer, heart and lung diseases, neurological problems, and asthma, [4]. Additionally, the resultant acid rain can damage and degrade properties and affect the economy of neighboring areas.

Considering the immense impacts of critical infrastructure facilities on the sustainability triple bottom line [6], it is crucial to develop a comprehensive plan throughout the life-cycle of these facilities to maintain their sustainable performance. While these impacts are mainly studied and the strategies to control them

Crystal River Nuclear Plant

Kewaunee Nuclear Station

San Onofre Nuclear Generating Station

Clinton Nuclear Plant

Figure 2.
Four critical infrastructures shutdown in the USA [8].

are established at the pre-design phase [7], due to a large number of internal and external factors (see **Figure 1**), all the potential effects of critical infrastructure on the sustainability cannot be predicted until the facility starts its operation.

The impacts of the critical infrastructure on the economy, immigration, ecosystem, and environment may take a direction different from the forecasted scenarios. Examples are the shutdown of four nuclear power plants in the USA (see **Figure 2** and **Table 2**) which wasted billions of dollars from the taxpayers. The unpredictable aspects of infrastructure are primarily because many external factors that can impact the sustainability performance of infrastructure are not predictable. Additionally, the external factors can affect the tolerance of the hosting environment in dealing with the negative impacts. Hence, the sustainability maintenance plans of infrastructure must be updated to effectively respond to the changes and decelerate the sustainability deterioration of critical infrastructure.

2. Best practices to maintain the sustainability of infrastructure projects

This section discusses the best practices to maintain the sustainability of infrastructure projects during their service life. These practices are explained in five categories as it can be seen in **Figure 3**. In each category, the recommended policies and practices are sub-categorized and discussed. These practices can be applied to any infrastructure facility or specific ones. Facility managers, decision-makers, and maintenance operation personnel must understand the necessity of investing in the implementation of these practices and motivate other personnel to a collaborate effort for maintaining a high quality of service from the facility as well as minimizing the negatives impacts of the infrastructure on the triple bottom lines of sustainability.

2.1 Reducing the negative impacts of infrastructure on people through effective maintenance

Depending on the size, type, and location of critical infrastructure facilities they can potentially affect large groups of people. These people can belong to one of the following categories.

1. People who work at the facility (full-time employees).

2. People who live in the vicinity of the facility (neighbors).

3. People who regularly commute to the facility (clients).

The sustainability maintenance measures to protect the health and well-being of people who belong to each of these categories can vary. There are comprehensive safety instructions to protect the people inside the facilities (including personnel and clients). The following practices in this section are primarily focused on protecting the people who live in the vicinity of an infrastructure facility (group 2). It is essential to keep in mind that people's health is the tops priority of sustainable development and it cannot be compromised or put at risk.

2.1.1 Maintaining community goals

While an infrastructure facility is developed to serve the public, it can be detrimental to the hosting community goals. Consider this case: A power plant that was established decades ago in the proximity of some small neighborhoods, initially led to the creation of job opportunities and a considerable amount of immigration to this area. The small neighboring communities who live near this facility were pleased about the foundation of this power plant as the value of their properties increased dramatically due to the economic growth. The facility administrators are now going to extend the facility's production. Locals are against this decision as they may believe that the air pollution caused by the power plant has increased in the last few years. This can be attributed to an increase in the production of the plant that can lead to more pollution. The power plant is now a burden on the community as the area has sufficiently developed and people feel that it is creating more problems for the community than contributing to the local development. The enlargement of the existing power plant threats the community goals by diminishing the quality of life and vitality of the locals. This example explains how a sustainable facility can turn to be detrimental to community goals and values.

The maintenance process creates the potential to identify and fix the gradual deviation of the infrastructure from the community expectations and goals. The community needs to notice that their concerns are considered and their voice is heard by the infrastructure's decision makers [9]. In order to meet the community's concerns people or their representatives must be involved in the decisions. Different tools can be used to communicate with the community and hearing their concerns. Survey questionnaires can particularly be helpful when the decision makers have to select between different requests by the public to figure out which options are more vigorously requested by the public. Communicating with stakeholders and inviting them to meetings to express their concerns about the issues caused by the facilities that can be mitigated through maintenance is another possible solution to engage people.

Another critical consideration in the assessment of the community issues and concerns and incorporating them into the maintenance plans is the conflict of

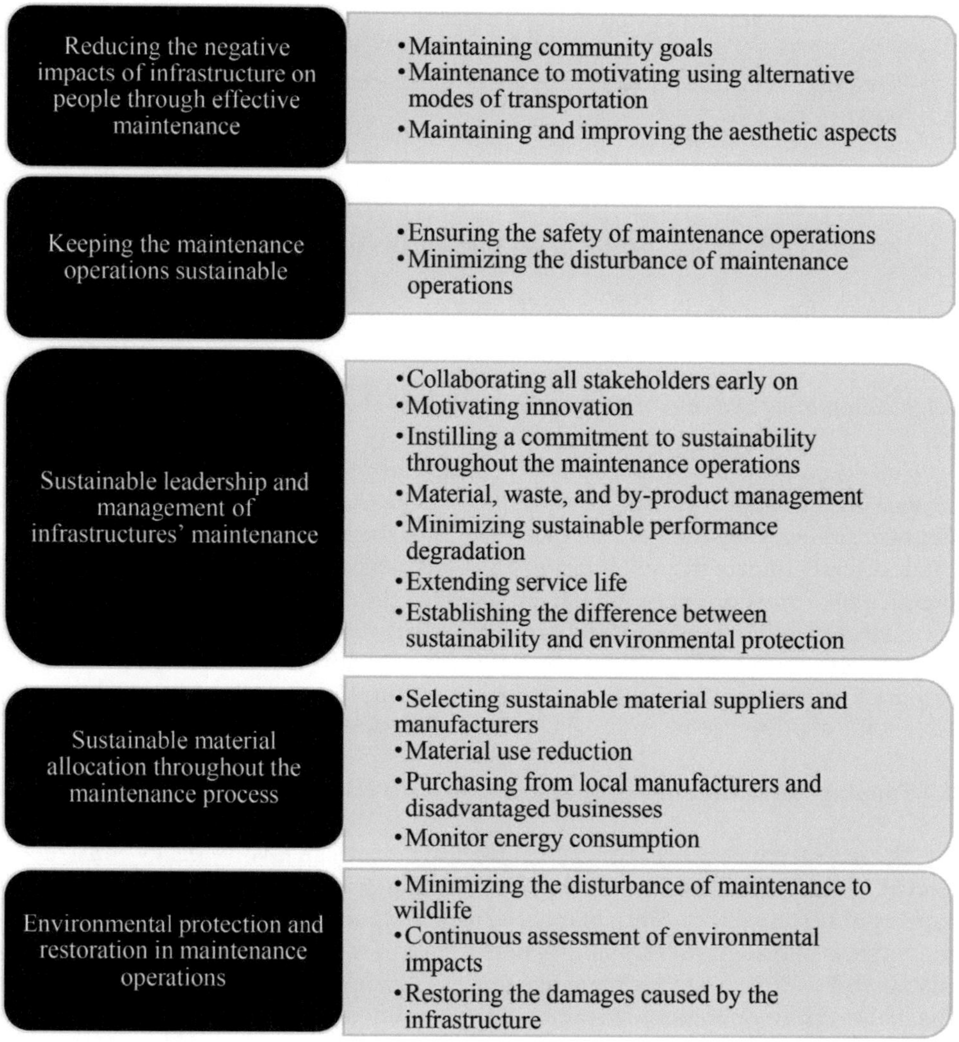

Figure 3.
Five categories of sustainability maintenance of critical infrastructures and their sub-categories.

interest between different communities [10]. This happens when one community benefits from a decision while another community finds it detrimental for its common goals. In these situations, hearing people's voice and their concerns can lead to informed decisions that more effectively address the concerns of the affected communities (**Figure 4**).

2.1.2 Maintenance to motivate using alternative modes of transportation

Many critical infrastructure facilities are extended to meet the growing demand of the hosting community for the type of service they provide. As a result of this extension, more people need to commute to these facilities as full-time employees or clients. The maintenance of critical infrastructure creates an opportunity for motivating people to use alternative modes of transportation (such as walking or using bicycles) to commute to the facility. This is particularly important for infrastructures with a high number of visitors and employees such as commercial facilities. Some suggested policies are listed below [11].

- Assigning priority parking for hybrid vehicles.
- Creating facilities to use bicycles (parking spaces for bikes and shower rooms for the drivers).
- Using gulf cars for the trips within the facility.
- Installing electricity charger in parking for electric vehicles.
- Improving access to public transit.

2.1.3 Maintaining and improving the esthetic aspects

Considering the magnitude of many critical infrastructures, and their long service life, the appearance of many of these facilities deteriorate, and if they are not maintained, they can become the eyesores of the hosting community. This can indirectly impact the neighboring areas and decline their property value. The maintenance must preserve the attractiveness of the community to live and work through the application of different measures that can enhance the appearance of the infrastructure facility. Some examples of performing this are using barrios that can mask an unpleasant view or renewing the paint, fixing the fences, and taking advantage of advertisements on old looking walls or other surfaces.

2.2 Keeping the maintenance operations sustainable

The maintenance process of critical infrastructure can involve heavy construction or lengthy operations with a lot of potential negative impacts on different aspects of sustainability. Since one of the primary purposes of the maintenance measures is improving sustainability performance, it is crucial not to compromise sustainability during the maintenance process. Some of the measures for keeping the maintenance operations sustainable are explained in the following.

2.2.1 Ensuring the safety of maintenance operations

Many infrastructure maintenance activities involve risks for the people who work in the project or who live in the project's proximity [12]. Hazards involved in construction or maintenance of infrastructure facilities could be even more than

Figure 4.
Best practices to maintain community goals throughout the service life an infrastructure facility.

typical (non-infrastructural) facilities as most of them are unique, extensive, and challenging to identify the potential risks during the maintenance operations. As an example replacement of cables of the Golden Gate Bridge requires working at the 746′ above the water level where rain and wind are common. At such condition safety measures to protect the laborer from falls is essential.

The hazards of maintenance operations can pose risks to the public too [12]. Trench collapses and scaffold collapses are some examples. Proper site fencing is one of the major preventive measures. The public can find watching the maintenance operations, such as replacing a segment of a street bridge, interesting. The crowd gathered around a maintenance site is a potential risk that must be handled. In many cases, the movement of large construction equipment, particularly when exiting the site, can cause the risk of accidents with other vehicles, buildings in congested urban areas, and pedestrians. There are comprehensive safety measures in each country to protect the public against these hazards. Any negligence or underestimating the importance of cautiously implementing safety measures must be eliminated throughout the entire maintenance operations.

2.2.2 Minimizing the disturbance of maintenance operations

Many maintenance measures can potentially create a lot of dust, noise, road closure, traffic congestion, odor, power outage, vibration, light pollution, and other problems to the community. It is essential to ensure that such disturbances are minimized. Some suggested policies are listed in **Figure 5**.

2.3 Sustainable leadership and management of infrastructures' maintenance

A large number of factors are involved in sustainability [13]. Examples are people and their health, well-being, and vitality; planet and the impacts of pollutions, energy consumption, and depleting resources on wildlife and habitat; and the economy with its connections to industrialization, and growing demand for more raw materials and disturbances in nature [14, 15]. This means that a great deal of interdisciplinary knowledge and expertise are required to make a sustainable decision. Additionally, every decision has its tradeoffs that need to be evaluated, and it can be challenging to decide what to compromise in the hope of gaining something that is more crucial to the public. The term public in this context is not only the people who might be impacted by that decision; it also spans to future generations [16]. The pressure of profitability almost always amplifies this complexity as it is a powerful incentive for decision-makers to accept the downfalls of a decision in the hope of economic growth.

While the concept of sustainability seems to be a straightforward approach for most people, due to its comprehensiveness and the conflicts of interests which it

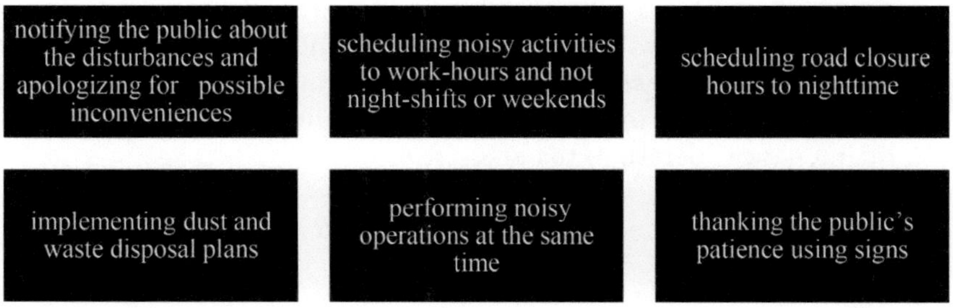

Figure 5.
Best practices to minimize the disturbance of maintenance operations for the neighbors.

inherits, it requires a lot of leadership and management. Some of the best practices of sustainable management of infrastructure are discussed in this section.

2.3.1 Involving all stakeholders from Early Stages

While the need for the collaboration of stakeholders in decisions seems to be obvious, the significance of their early collaboration might not be sufficiently noticed. People, particularly, must be involved before decisions are made to ensure the concerns of the community are adequately considered in the decision making process [17]. This can be done through the managers' determination to integrate the public's demand and concerns in the decisions and involve experts from different disciplines to come up with optimized decisions. It is crucial to keep in mind that maintenance provides an opportunity to adjust and change an existing infrastructure based on the actual issues observed during its operation.

The public's engagement should not be considered an additional challenge for the project. Reflecting the public's input in the project is an essential factor in coming up with the decisions in which the public's concerns are considered, and negative impacts on the hosting community are minimized. This requires building a relationship with the key stakeholders. Although in some regions there are regulations that enforce the involvement of the public, the sustainability goals go beyond fulfilling the minimum required standards and seek for active engagement, transparency in notifying the public about potential impacts and incorporating them in the process of decision making.

Managers must facilitate communications using all the possible tools. While many maintenance projects are composed of the involvement of multiple independent entities who are primarily focused on delivering their portion of work, managers must minimize isolated thinking and establish a collaborative approach in which impacts on sustainability can be easily communicated between different project players. Similar to integrated project delivery in construction, the maintenance operations should also attempt to reach to an optimized collaborative environment from the start point. This approach is in contrast with the traditional methods in which project parties prefer to work separately. The managers' determination and attempt to change the independent work environment to a collaborative work environment is essential in the success of sustainability plans in an infrastructure maintenance project.

2.3.2 Motivating innovation

The maintenance management team must promote any changes in the existing maintenance strategies that can potentially improve sustainability. An effective change is utilizing new technologies for data collection. The amount and quality of data about infrastructure are closely linked with the quality of maintenance. Management must establish a culture of "change for the better" in the maintenance team to seek for any modifications in the existing maintenance policies that can lead to improvement. This improvement can be about cutting the maintenance costs, improving the quality of the infrastructure serviceability, or enhancing the sustainability of maintenance operations to name a few.

2.3.3 Instilling a commitment to sustainability throughout the maintenance operations

The management's commitment to sustainability is pivotal in promoting or discouraging sustainable approaches among all team members [18]. The managers must make sure that the team has a sufficient understanding of the threats of all the maintenance operations to sustainability and work effectively with them to minimize these threats. This commitment must be observable in the design, schedule, field activities, and inspections. The team must be aware that the management has no willingness in compromising sustainability for less critical goals such as performing the operations more quickly, or cutting the costs unless they can be justified in a sustainable-thinking context. Sustainability workshops before beginning the maintenance operations are often needed to transfer the determination of the company to contribute to sustainability to the project players as well as teaching them the best practices that trigger sustainability.

Commitment to sustainability must be established as a subject that is as important as safety. While this may seem exaggerated for some contractors, they can understand it better considering that unsustainable methods can potentially impact the health or well-being of communities at local, national, or even international levels, while unsafe operations mostly pose risks to the staff on the job site or the neighbors only. Moreover, since protecting the health of human beings is an indispensable element of sustainable development [14], improving the job site safety perfectly lines up with the goals of sustainability [19]. Combining safety and sustainability workshops are expected to contribute to instilling a crucial need for sustainability among the maintenance team.

Some of the potential impacts of the maintenance operations on the hosting environment can only be noticed during the operations. Therefore, throughout the operations, the management team is responsible for supporting any decision or activity that can enhance the sustainability of the operations and achieve a level of sustainability that is beyond the original plans.

2.3.4 Material, waste, and by-product management

Although material, waste and by-product handling is a team effort, the management role in leading the team is substantial. Material handling aims at maximizing the efficiency of the construction and maintenance materials. There are various methods and best practices to implement it. Some highly recommended tools are briefly explained in the following.

1. Just-in-time (JIT) procurement of materials which attempts to minimize the need for storing materials and bringing them to the site when they can be almost immediately used in the maintenance operations. The advantages of JIT material procurement is shown in **Figure 6** [20].

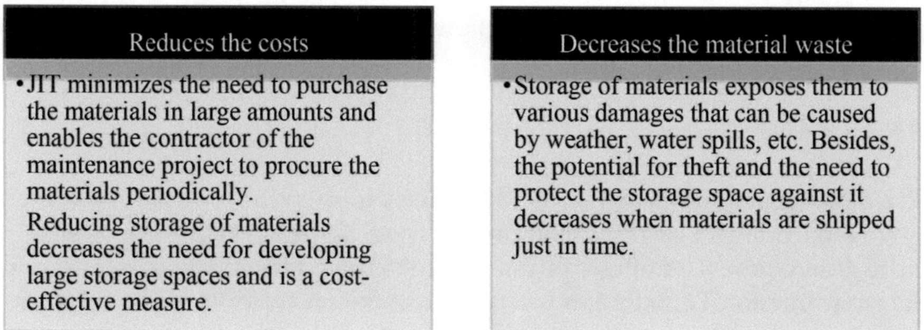

Figure 6.
Advantages of shipping the material to the facility Just-In-Time.

2. Waste management workshop for the team. There are a lot of potentials for effective management of waste and by-products in a maintenance project. Project-specific opportunities for reduction of waste, reuse, recycle, convert, or compost must be identified by the managers or their representatives. Laborers should receive adequate instructions about the firm's waste management policies and be motivated to contribute to it.

2.3.5 Minimizing sustainable performance degradation

It is often observed that sustainability concerns are usually centered to the facility's design and construction phases (rather than the operation phase) and once an infrastructure facility starts operation, the focus of its managers is maintaining its level of service at the expected quality and. This explains why the negative impacts of a facility on sustainability becomes more noticeable in the course of its service life. Another reason for sustainable performance degradation are populist managers who make decisions to pretend their attention to sustainability, while they refuse to accept the sustainability costs that people or media will not necessarily notice or appreciate.

Maintenance operations are primarily focused on replacements that can improve the serviceability or durability of a facility. Less attention is commonly paid to fixing the sustainability issues. Although infrastructure projects are often designed with careful examination of the potential impacts of the facility, due to the dynamic nature of the surrounding environment and other modifications in the facility in the course of its service life, many sustainability issues may arise when the facility is in operation. Managers have a critical responsibility to continuously measure and mitigate the sustainability threats of the project. Maintenance for sustainability can encompass low-cost activities such as eliminating the noise caused by displacement of a utility hole cover to more intensive maintenance such as restoring the vegetation affected by the pollution. The target is to maintain the infrastructure surrounding in such a way that the facility's footprint is almost none beyond its shell. This requires a specific focus in maintenance that is supported and pursued by the managers.

What makes maintaining sustainable performance more challenging are the changes in the management team. Different managers have different priorities in mind that can impact their determination for maintaining the facility's sustainability. A useful tool that can prevent underestimating sustainability in maintenance policies is developing a comprehensive long-term plan for sustainable performance that can be pursued by future managers. On the other hand, relying on old approaches inherited from previous managers is not helpful as the project's impacts, and the demands for environmental protection can change. The managers must, therefore, seek solutions for improvement while staying committed to the long term plans of sustainable maintenance.

2.3.6 Extending service life

The development of an infrastructure facility is often the product of the monetary contribution of a large number of taxpayers. Some of the infrastructure facilities also require a lot of compromises due to their adverse impacts, particularly on the environment. Therefore, it is crucial to make every effort to extend infrastructures' service life as much as possible.

A great deal of a project's service life depends on the quality of its design. Flexible design enables the project to respond to the variations in public's demand from the facility, and the dynamic changes of the facility as well as the external factors it interacts with. The design must allow reconfiguration as well as refurbishment of the facility and minimize the need for demolition or replacements. Additionally, the design must establish a high resiliency in the facility to survive against the external pressures. Though, even with a perfect design, the facilities will gradually start to deteriorate. As can be seen in **Figure 7** [21], two main approaches can be applied to extend the service life of infrastructures, maintenance and rehabilitation. Both approaches primarily depend on budget and project condition. Maintenance is done in shorter intervals, and higher frequencies improve the asset condition [22]. When regular maintenance activities cannot effectively improve the asset condition, a rehabilitation or renewal activity must be pursued. This requires more time, budget, and effort but, as they can be seen in the graph, can more significantly improve the infrastructure condition. In this figure, the distance between the two parallel diagonal dotted lines on the x-axis indicates the added time to the service life as a result of maintenance and rehabilitation.

Facility managers must be mindful about a load of service on the facility even though it is below the capacity. A higher load of operation is linked with quicker deterioration. The manager should keep close attention to avoid over-production to lower the unnecessary load to the infrastructure. Setting various inspection intervals depending on the frequency of the need for repairs in different components of a facility is another helpful approach. Long inspection periods for all components is detrimental to effective maintenance which is geared towards extending the facility's service life and must be avoided.

2.3.7 Establishing the difference between sustainability and environmental protection

Although environmental protection is a critical goal of sustainable development, it is essential to understand that sustainability goes beyond environmental protection and involves social and economic goals. The establishment of this understanding is the responsibility of the managers. This can be challenging because understanding the potential threats to communities can be more complicated than figuring out the environmental threats. Some maintenance decisions may not have negative environmental impacts, but they may be detrimental to the society or economy.

Here is an example of the necessity of paying attention to the social aspects of sustainability besides the environmental aspects. Consider a coal plant with a private access road that is far enough from a residential neighborhood. While

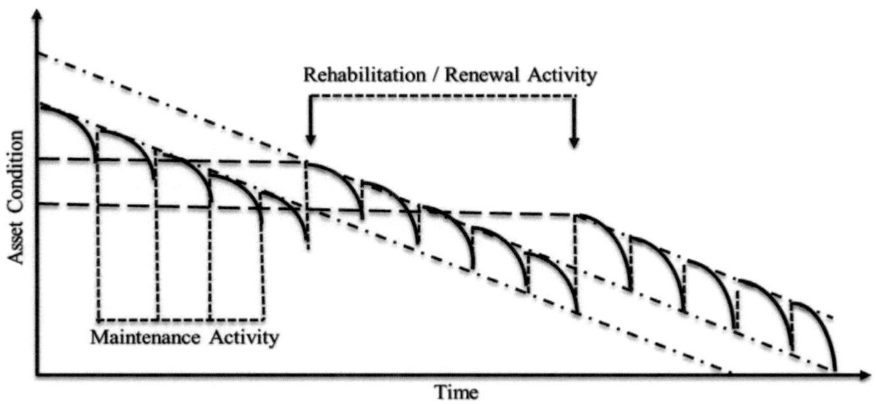

Figure 7.
Extending infrastructure service life through maintenance and rehabilitation/renewal.

developing the maintenance plan, the plant's manager suggests relocating the access road to the neighborhood nearby that has a low population can save the travel time for the trucks that supply the materials for the plant. The decision seems to make sense from the environmental aspect as it saves a few minutes of travel time and can potentially lead to a reduction in the pollution caused by the trucks in the long run. However, it will impact the people living in the small neighborhood on a daily basis. Therefore, if environmental protection is the sole criteria for making a sustainable decision the social and economic aspects of sustainability can be compromised, and this defeats the purpose of sustainable development.

It is essential to establish equity in pursuing sustainability [23] Equality in sustainability can be defined as being impartial to the communities while making decisions that can impact the triple-bottom-line of sustainability in neighboring areas. Lopsided progress in sustainability programs by putting one community prior to another is in contrast with sustainability goals. Bearing in mind that trade-offs are inherited in many decisions tied to development, the community facing the consequences of a decision may be different from the community that benefits from it. Keeping the balance between neighboring communities in benefiting from the opportunities promotes sustainable development.

2.4 Sustainable material allocation throughout the maintenance process

Growing concerns about depletion or shortage of resources, in addition to increasing demands for more materials, explains the necessity of efficient resource allocation [24]. Depending on the type of infrastructure and the depth of the maintenance operations, the volume of required materials for the maintenance can vary from small amounts to enormous quantities. For example, tons of concrete can be required in the replacement of a bridge segment. The way these materials are selected, shipped, and used in the maintenance operations have impacts on the triple bottom lines of sustainability. According to the World Counts, on average everyone uses 16 kilos of resources extracted from the earth every day, and for people in the western world, this number is much higher—up to 57 kilos of newly-mined minerals per day [25]. Based on the United States Geological Survey (USGC) the U.S. apparent consumption of raw materials at the beginning of the century has

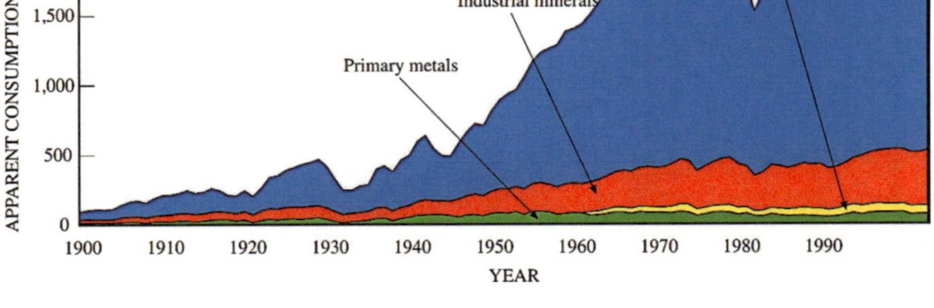

Figure 8.
U.S. apparent consumption of raw materials (courtesy of USGS).

increased more than six times from 1940s (see **Figure 8**). The infrastructures are massive consumers of resources worldwide. For example, a total of approximately 1.5 billion tons of aggregates, 35 million tons of asphalt, 48 million tons of cement, and 6 million tons of steel is in place in interstate highways of the USA [25].

In addition to the depletion of resources, transportation and shipping of raw materials can affect the environment in multiple direct or indirect ways. The massive amount of construction materials that were mentioned in the example above must be transported to sites through thousands of trips by heavy equipment that contributes to air, water, and noise pollution as well as fuel consumption. The indirect effects of trips to supply the materials for the infrastructure can indirectly contribute to global warming, climate change, and threaten the health of human beings. Additionally, continuous traffic of heavy equipment to an infrastructure (such as plants and refineries) will have impacts of the property value of the surrounding neighborhoods and affect the economy bottom-line of sustainability.

These examples indicate that any improvement in the way the infrastructure materials are procured is expected to significantly contribute to sustainable development.

2.4.1 Selecting sustainable material suppliers and manufacturers

Purchase of materials is commonly done through bidding and choosing the least expensive bidder that meets the expected quality. Less attention is paid to how sustainably the supplier or manufacturer procures the material. Selecting the material supplier is a multi-criteria decision [26, 27]. Selection criteria of material supplier or manufacturer must modify and include the following criteria (**Figure 9**).

- Extraction methods used by the supplier. This explains the policies the manufacturer uses in extraction to maintain the balance of resources in nature and what measures they utilize to minimize the negative impacts on the ecosystem. There are six principles of green extraction [28]: 1) using renewable plant resources, 2) using alternative solvents, 3) reducing energy consumption by energy recovery, 4) converting wastes to co-products, 5) reducing unit operations, and 6) maximizing non-denatured and biodegradable uncontaminated extracts.

- Green processing of the raw materials. This defines how sustainably the extracted materials are processed by the manufacturer. There are multiple criteria to measure this. Some of these criteria are as follows: (1) Compliance to physical safety (safety measure regarding flammability, explosivity, corrosion, oxidation, and radioactivity), (2) compliance to air pollution reduction (meeting the allowable impacts on air, water, and soil), and (3) sustainable waste management (waste generation reduction, green recycling methods, recyclability of products).

- Green shipment. This defines the measures the manufacturer utilizes to minimize the pollution generated in the shipping process. Some criteria to measure this are as follows: (1) using low polluting vehicles and (2) efficient shipment design (to minimize the number of trips).

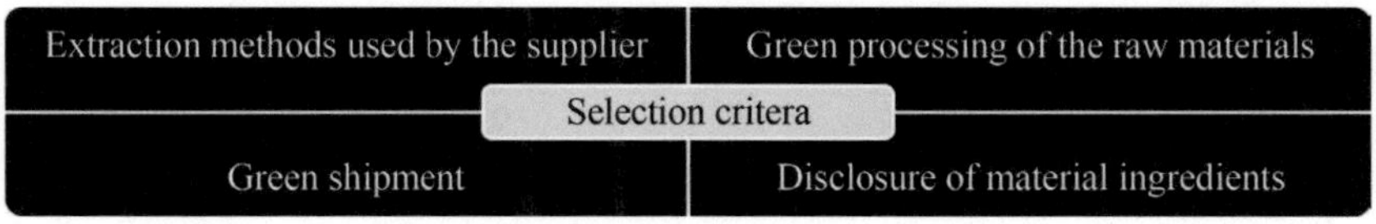

Figure 9.
Selection criteria of material supplier or manufacturer.

- Disclosure of material ingredients. The supplier must disclose the life-cycle information of the chemicals they have used in the product. This is more important for the infrastructure facilities that are in the proximity of children or other sensitive groups.

2.4.2 Material use reduction

The maintenance operation must seek for minimizing the need for raw materials. This can be achieved without compromising the quality of maintenance if certain measures are taken collaboratively. Some of these measured are explained in the following.

- Using recycled materials. Using recycled, reused, and renewable materials reduces the demand for raw resources and contributes to preserving them for future generations. Maintenance operations that involve reconstruction of segments have high potentials to reuse certain elements of the existing part that is demolished. This requires an assessment of the opportunities to reuse and planning it before the demolition starts. Additionally, repair policies instead of replacement, particularly for equipment pieces, can significantly contribute to saving the embodied energy that has been used to manufacture, ship, and install the existing equipment. However, in all cases, the quality and performance criteria must not be compromised.

2.4.3 Purchasing from local manufacturers and disadvantaged businesses

Procurement of materials from local suppliers lines up with sustainability goals in two ways: (1) by supporting the local businesses and contributing to economic growth and the welfare of the local community and (2) by reducing the negative impacts of shipment of materials on sustainability (air pollution caused by the shipping vehicles, costs, and risks of road accidents).

In addition to local suppliers, the disadvantaged businesses must be supported while selecting the suppliers, or manufacturers. In the U.S. public projects, there is commonly a specified minimum percentage of purchases that must be obtained from the disadvantaged businesses. Examples of these businesses are African Americans, Hispanics, Native Americans, Asian-Pacific and Subcontinent Asian Americans, and women-owned companies. Considering that most infrastructure facilities are public, maintenance purchases that are often in smaller amounts compared with new construction purchases creates an opportunity to support disadvantaged business although these businesses may not afford to be the lowest bidder.

2.4.4 Monitor energy consumption

One of the purposes of infrastructure maintenance must be maintaining the infrastructure's negative impacts within the decided limits throughout the service life of the facility. For an operating infrastructure facility, a large amount of these impacts depend on energy consumption rates. The energy performance of infrastructure facilities is expected to decline throughout the years of service. Maintenance must fix the degradation of energy performance. This depends on the quality of energy monitoring that enables the facility managers to identify were the energy losses are occurring and fix them.

2.5 Environmental protection and restoration in maintenance operations

Although infrastructure projects are often designed and constructed by studying and minimizing their potential impacts on the environment, due to a variety of mostly unpredictable causes, these facilities can cause different damages to the neighboring areas. It is essential to include the protection and restoration of such damages in the maintenance policies. The degradation of nature happens gradually, and the damages can be difficult to notice without inspections.

In Addition to restoration, the causes of the damage to nature need to be studied, and preservation from further damage must be added to the maintenance programs of the infrastructure. For instance, if soil pollution has been reported, different possible scenarios of the causes of pollution need to be considered. The common practice in such cases is relying on the most obvious cause for the damage while further assessment can reveal some causes that may seem unlikely. The comprehensiveness of the assessment has a central role in the quality of maintenance programs.

Infrastructure facilities affect the surrounding environment in various ways. Although some compromises about the negative environmental impacts are inevitable, taking certain measures can significantly reduce these impacts. The majority of the environmental protection activities are related to the design and construction phases; however, the maintenance policies of infrastructures can contribute to effectively preserve its surrounding environment. What follows are some of the methods to enhance environmental protection while the facility is operating.

2.5.1 Minimizing the disturbance of maintenance to wildlife

Depending on the facility's location, the maintenance operations can disturb wildlife in different ways. While a lot of consideration is commonly given to minimizing the disturbance to wildlife during the construction process, the risks of maintenance operations on the environment are, in many cases, underestimated and this can create severe threats for the habitats and wildlife. These risks are increased when wildlife is exposed to them for longer durations. This exposure can even impact migration, feeding, and breeding patterns. For instance, using vibrators in maintenance operations of a dam can affect the fish reproduction in adjacent waters.

EPA provides comprehensive guidelines and requirements [29] to ensure that maintenance operations' disturbances to wildlife are minimized. Some primary measures in doing so are listed in the following:

- Identifying vulnerable species in the project vicinity (including species sensitive to noises or vibrations, species that need large habitats, and species with low reproductive pedigree).

- Maximizing off-site operations (instead of performing the tasks on an eco-sensitive site).

- Minimizing different pollution types (air, water, soil, noise, and light pollution). Depending on the type each of the mentioned pollutions requires taking multiple measures for prevention.

- Establishing stringent fire-protection measures and regulations (for transferring and storing flammable materials, equipment fuel procurement, and smoking on the site).

- Restoring the changes on the natural habitat during after the operations (removing temporary equipment access roads by replacing the polluted or compacted soil with the natural soil of the area, restoring the damages vegetation and views.

2.5.2 Continuous assessment of environmental impacts

Depending on the type, size, and location of the infrastructure it can affect different elements of the natural world including habitats and species. The interactions between the facility and its surrounding environment can vary over time because both the infrastructure and the environment are dynamic systems with altering elements that interact with the elements of one another. For instance, the amount of pollution infrastructure generates can change based on its production. On the other hand, the tolerance of the adjacent environment can also modify based on the severity and frequency of rainfalls, the intensity of winds, and temperature inversion in different seasons or years. Another example is the tolerance of water bodies to pollutants which decrease as they are exposed to more contaminants. In this situation, the infrastructure may have to reduce its production to maintain the water quality before other preventive measures can be implemented.

The environmental assessment methods, required tools, intervals of measurement, and data analysis methods depend on the type of environment that is in the proximity of the infrastructure. Figuring out the pattern of the repetitive alterations in the reactions between the infrastructure and the responses from the environment can significantly improve the preventive measures.

2.5.3 Restoring the damages caused by the infrastructure

Paying adequate attention to preventive measures is essential in the management of infrastructure. Though, considering the complexities and unpredictability involved during long years of service of infrastructures some damages to the adjacent environment are inevitable. However, these damages can be minimized by executing restorative measures. Depending on the type of damage, a wide variety of options exist for restoration purposes. It is crucial to keep in mind that some of these damages cannot be readily observable and they require a more in-depth examination of the area and evaluation of the facility's impacts.

Restorative measures are focused on bringing the adjacent environment back to the condition it had before the infrastructure was developed, as much as possible. These measures are more critical when the damages affect residential areas, wildlife, habitat, rare species, greenfields, wetlands, prime farmlands, and historical sites.

The restorative measure can have different types depending on the type of damages they try to restore. Some of the main types are the following:

- Emergency restoration. This is when the damage should be taken care of immediately. For examples the leakage of pollutant fluids. The facility must develop a plan with the collaboration of different organizations to react quickly when such damages occur. In the case above, after the leakage is stopped, immediate action is required to remove the spilled fluid before the contamination transfers to lower layers of soil or a water body.

- Restoring the damages of a fire or smoke. One of the most critical impacts of infrastructure on the environment is increasing the risk of fires. Adequate preventive measures must be taken to mitigate the risk of fire transfer from the facility to adjacent areas. In addition to these measures, in case a fire happens, the damages it has caused must be restored by recreating the affected vegetation or removing the soot, water used in extinguishing the fire), and odor. In cases that restoration is not possible, the surface area and types of damages must be measured, and an equal amount of vegetation must be planted in an alternative site.

- Restoring disturbed soil. Natural soils are affected by the infrastructure in two major ways: (1) oil and other pollutants' spill and (2) compaction of soil as a result of temporary traffic of heavy equipment. Some maintenance operations utilize heavy equipment pieces that pollute and compact the natural soil. The facility managers tend to keep the temporary access roads in case a need arises in the future. This approach must be avoided as it unnecessarily expands the footprint of the facility.

In almost all restoration activities, promptness goes a long way and delaying the fixes can lead to irreparable damage. The key to prompt restoration is thinking ahead and having the required plans ready to step in when the need for restoration urges.

3. Conclusion

Critical infrastructures have significant adverse impacts on sustainability that, if not handled, can defeat their purpose of serving to the public. Although a lot of attention is paid to develop a critical infrastructure with a minimal footprint, once a facility is operating less consideration is paid to evaluate how it interacts with its surrounding environment or the host community. It was mentioned that the dynamic nature of the infrastructure and its hosting environment causes a lot of alterations in how the facility impacts its surroundings. Examples such as the shutdown of nuclear power plants in the USA indicate that in many cased the planned trajectory of infrastructure may not be pursued due to various internal and external factors that affect infrastructures. Without proper maintenance of sustainability performance, the negative impacts of infrastructures will gradually accumulate and may lead to irreversible damages. Continuous data collection, inspections, and measurement of the environmental impacts are essential to identify how the facility's performance must be modified to maintain its negative impacts within the desired limits.

Sustainability maintenance is an ongoing process throughout the service life of infrastructure to maintain its negative impacts within the desired limits and ideally enhance its sustainability performance. It was mentioned that regular maintenance operations provide a significant opportunity to mitigate the negative sustainability impacts of infrastructure. Unlike the design phase in which the sustainability threats are predicted, while performing the maintenance operation, most of the sustainability-related issues are precisely identified.

The significant roles of considering the hosting community's concerns, managers' determination to sustainable performance, sustainable handling of materials, sustainable maintenance operations, and continuous restoration of damages caused by the infrastructure were highlighted.

Author details

Mohammadsoroush Tafazzoli
Washington State University, Pullman, WA, United States of America

*Address all correspondence to: tafazzoli@wus.edu

References

[1] Baillie C. Synthesis lectures on engineers, technology, and society. In: Engineers Within a Local and Global Society. San Rafael: Morgan & Claypool Publishers; 2006

[2] Palei T. Assessing the impact of infrastructure on economic growth and global competitiveness. Procedia Economics and Finance. 2015;**23**:168-175

[3] National Research Council. Sustainable Critical Infrastructure Systems: A Framework for Meeting 21st Century Imperatives: Report of a Workshop. National Academies Press; 2009

[4] Unions of Concerned Scientists, Science for a healthy planet and safer world. Available from: https://www.ucsusa.org/clean-energy/coal-and-other-fossil-fuels/coal-air-pollution [Accessed: 27 February 2019]

[5] United States Environmental Protection Agency, 2014 National Emissions Inventory (NEI) Data. Available from: https://www.epa.gov/air-emissions-inventories/2014-national-emissions-inventory-nei-data [Accessed: 27 February 2019]

[6] Little RG. Tending the infrastructure commons: Ensuring the sustainability of our vital public systems. Structure and Infrastructure Engineering. 2005;**1**(4):263-270

[7] Biggs CTB, Ryan CJR, Wiseman JR. Distributed Systems: A design model for sustainable and resilient infrastructure. Melbourne: Australia; 2010

[8] Power Engineering., U.S. Nuclear Power Plants Already Closed or Closing. Available from: https://www.power-eng.com/articles/slideshow/2016/08/u-s-nuclear-power-plants-already-closed-or-closing, [Accessed: 27 February 2019]

[9] Li TH, Ng ST, Skitmore M. Conflict or consensus: An investigation of stakeholder concerns during the participation process of major infrastructure and construction projects in Hong Kong. Habitat International. 2012;**36**(2):333-342

[10] Eden H, Scharff E, Hornecker E. Multilevel design and role play: Experiences in assessing support for neighborhood participation in design. In: Proceedings of the 4th Conference on Designing Interactive Systems: Processes, Practices, Methods, and Techniques. ACM; 2002. pp. 387-392

[11] Ogilvie D, Egan M, Hamilton V, Petticrew M. Promoting walking and cycling as an alternative to using cars: Systematic review. BMJ. 2004;**329**(7469):763

[12] Hamid AR, Yusof WZ, Singh BS. Hazards at Construction Sites. Universiti Teknologi Malaysia; 2003

[13] Rametsteiner E, Pülzl H, Alkan-Olsson J, Frederiksen P. Sustainability indicator development—Science or political negotiation? Ecological Indicators. 2011;**11**(1):61-70

[14] Robert KW, Parris TM, Leiserowitz AA. What is sustainable development? Goals, indicators, values, and practice. Environment: Science and Policy for Sustainable Development. 2005;**47**(3):8-21

[15] Bossel H. Indicators for Sustainable Development: Theory, Method, Applications. International Institute for Sustainable Development. Winnipeg; 1999. p. 138

[16] Beckerman W. Sustainable development and our obligations to future generations. UK: Oxford University Press; 1999

[17] Fadeeva Z. Promise of sustainability collaboration—potential fulfilled? Journal of Cleaner Production. 2005;**13**(2):165-174

[18] Reed R, Lemak DJ, Mero NP. Total quality management and sustainable competitive advantage. Journal of Quality Management. 2000;**5**(1):5-26

[19] Rajendran S, Gambatese JA. Development and initial validation of sustainable construction safety and health rating system. Journal of Construction Engineering and Management. 2009;**135**(10):1067-1075

[20] Pheng LS, Meng CY. Managing Productivity in Construction: JIT Operations and Measurements. Routledge; USA: 2018

[21] Cagle RF. Infrastructure asset management: An emerging direction. AACE International Transactions. 2003;**1**:PM21

[22] Tafazzoli M. Strategizing sustainable infrastructure asset management in developing countries. In: International Conference on Sustainable Infrastructure. 2017. pp. 375-387

[23] Tanguay GA, Rajaonson J, Lefebvre JF, Lanoie P. Measuring the sustainability of cities: An analysis of the use of local indicators. Ecological Indicators. 2010;**10**(2):407-418

[24] Tafazzoli M. A method to measure material-use efficiency in construction projects. In: The Proceedings of the 52nd Annual International Conference of Associated School of Construction. Provo, Utah; 2016

[25] The World Counts. It's Minded for Your Sake, Available from: http://www.theworldcounts.com/counters/environmental_effect_of_mining/depletion_of_natural_resources_statistics [Accessed: 28 February 2019]

[26] Jia P, Govindan K, Choi TM, Rajendran S. Supplier selection problems in fashion business operations with sustainability considerations. Sustainability. 2015;7(2):1603-1619

[27] Chemat F, Vian MA, Cravotto G. Green extraction of natural products: Concept and principles. International Journal of Molecular Sciences. 2012;**13**(7):8615-8627

[28] Walker H, Preuss L. Fostering sustainability through sourcing from small businesses: Public sector perspectives. Journal of Cleaner Production. 2008;16(15):1600-1609

[29] Thornton RD. Searching for consensus and predictability: Habitat conservation planning under the endangered species act of 1973. Environmental Law. 1991;**21**:605

Chapter 3

Modeling Resilience in Electrical Distribution Networks

Alberto Tofani, Gregorio D'Agostino, Antonio Di Pietro, Sonia Giovinazzi, Luigi La Porta, Giacomo Parmendola, Maurizio Pollino and Vittorio Rosato

Abstract

Electrical distribution networks deliver a fundamental service to citizens. However, they are still highly vulnerable to natural hazards as well as to cyberattacks; therefore, additional commitment and investments are needed to foster their resilience. Toward that, this paper presents and proposes the use of a complex simulation model, called reconfiguration simulator (RecSIM), enabling to evaluate the effectiveness of resilience enhancement strategies for electric distribution networks and the required resources to implement them. The focus is, in particular, on one specific attribute of resilience, namely, the readiness, i.e., the promptness and efficiency to recover the service functionality after a crisis event by managing and deploying the available resources rapidly and effectively. RecSIM allows estimating how and to what extent technological, topological, and management issues might improve electrical distribution networks' functionality after the occurrence of accidental faults, accounting for interdependency issues and reconfiguration possibilities. The viability of implementing RecSIM on a real and large urban network is showcased in the paper with reference to the study case of the electrical distribution network (EDN) of Rome city.

Keywords: electrical distribution network, resilience metrics, interdependence, outage recovery, vulnerability

1. Introduction

Electrical distribution network (EDN) delivers a fundamental service to citizens. Unfortunately they are still very vulnerable to natural hazards as well as to cyberattacks; both can affect electricity infrastructures, leading to power outages that might distress and delay the recovery of the impacted communities. In Europe, for example, adverse space weather, riverine floods, and earthquakes are recognized to be the prevalent hazards with high potential for disrupting the functions of the power grid. While high-voltage overhead transmission systems proved to be robust to earthquake hazard, earthquake-induced ground motion was recognized to cause inertial damage to electric distribution system, in particular, to heavy equipment, such as generators and transformers, and brittle items, such as ceramics, as well as to the building housing the substations; earthquake-induced ground failure and soil liquefaction were identified as one the main causes of damage to buried electric infrastructure components [1–3]. The time required for restoring power supply following earthquakes was seen to range from few hours to months (being more frequently in the range from 1 to 4 days) depending on the repair capabilities (e.g., availability of man power, machinery, and spare material) and on the level of access to damaged facilities, possibly delayed by damages to the road network and/or by traffic congestion [1].

As far as adverse meteorological conditions are concerned, both the transmission and distribution systems have been adversely affected by water bomb causing flooding, extreme snowfall or windstorm, and overheating [1]. As an example, high-voltage overhead lines might be subjected to failure due to ice sleeves on conductors during snowfalls; medium-voltage overhead lines might be subjected to failure due to fall of trees during windstorms, while overheating can cause catastrophic failure of underground cables [4, 5]. As an example, a clamorous case occurred in Auckland, New Zealand, in 1998 that involved the failure of four major underground cables due to overheating in the summer period. The failure of the underground cables kicked off a 5-week-long power outage across the central city and caused an estimated long-term economic impact equivalent to 0.1–0.3% New Zealand's gross domestic product.

From the few facts mentioned above, it is clear that additional commitment and investments would be worthwhile, if not needed, to foster the resilience of the EDNs.

EDN resilience can be pursued steadily before, during, and after crisis situations by putting in place, in an integrated and balanced way, various actions aimed at increasing the *robustness* of the network components; the *redundancy* of the system; the *resourcefulness*, i.e., availability of resources (such as backup systems, human and material resources); and the *readiness*, i.e., the promptness and efficiency to recover the service functionality after a crisis event by managing and deploying the available resources rapidly and effectively [6].

The works presented in this paper focuses on the resilience enhancement *after crisis events*, with particular emphasis on the factors that might increase the *readiness*.

A further aspect examined by this work is the interdependency of EDNs with other critical infrastructures (CIs) and the implication that this has on the resilience of EDNs. EDNs are, in fact, essential for the functionality of other services such as water, telecommunications (tlc), roads, and other public services; on the other hand, EDNs depend on other critical infrastructures to deliver their service. In particular, EDNs are highly dependent on telecommunication that provides telecontrol functionality to EDN, to such an extent that it is fair to assume that electrical and telecommunication networks do represent a unique, connected *system of systems* whose control, protection, and management should be performed as if it was a unique system.

The paper is organized as follows. Section 2 presents relevant works related to existing methods for the resilience assessment of EDNs. Section 3 contains a description of the abstract model representing the topology and the constitutive elements of a large EDN. Section 4 identifies metrics for assessing the resilience of EDNs in terms of induced service impacts after different kinds of perturbations. Finally, Section 6 presents the implementation on the case study of Rome, Italy.

2. Related work

All definitions of resilience point to quantify a *dynamic, adaptive property of a system* (or of a *system of systems*) expressing its ability to withstand perturbations and to recover, rapidly and effectively, to equilibrium condition as similar as possible to that prior to perturbation [6–9]. When dealing with a technological system, the property of being "adaptive" inevitably leads to think of a number of factors influencing the way the system might adapt other than its mere technological qualities (such as robustness, technological update of the components, etc.): these factors might include risk awareness and preparedness, to ordinary and emergency management capabilities, in general to management skills which must support the technological and the design quality of the network. Moreover, in the case of a *system of systems*, the resilience of a system also depends on the degree of resilience of the other systems whose services should be available for the recovery process and on the level of dependency that is between them. The nowadays emphasis on the resilience property of technological systems is a direct symptom of the increased awareness that networks' functional dependency is one of the major issues that must be considered for improving CI protection and, as such, should be always appropriately considered when dealing with modeling and simulation activities of these systems. In particular, in Europe different resilience assessment and management methods as well as new approaches and guidelines are proposed within interesting EU projects. The project SMR [10] proposes the guidelines and system dynamic modeling and simulation techniques to increase the resilience of cities, whereas the IMPROVER project [11] is more focused on CI. The project DARWIN [12] is focused on improving responses to expected and unexpected crises affecting critical societal structures during natural disasters (e.g., flooding, earthquakes) and man-made disasters (e.g., cyberattacks). To achieve this, DARWIN developed resilience management guidelines aimed at critical infrastructure managers, crisis and emergency response managers, service providers, first responders, and policy makers. Other interesting EU project results can be found in [13–15]. The main objective of these projects is the proposal of European Resilience Management Guideline (ERMG) frameworks to drive decision and policy makers, local governments, and CI operators toward more resilient cities, societies, and infrastructures. ENEA has proposed CIPCast a framework for the resilience evaluation of a specific area that is compliant with the general guidelines proposed, for example, in [10]. CIPCast allows geographical information system, GIS-based risk assessment, and situational awareness through the continuous acquisition of different kinds of data from the field (e.g., weather forecast, infrastructure network status). Furthermore, CIPCast allows the assessment of the impacts and consequences of possible damage scenarios due to the prediction of natural hazards (such as heavy rain, flash floods, earthquakes) on the infrastructure networks and services and on the affected communities [17, 18]. The present work describes RecSIM [19], a specific module of CIPCast allowing the operational resilience assessment of electrical distribution grids. Indeed, there is an increasing demand for resilience framework assessment of power grids due to the fact that electrical power grids are recognized as critical lifelines that have to cope with different threads including extreme natural disasters and man-made attacks [20–25]. An extensive review of the existing metric system and evaluation methodologies, as well as a quantitative framework for power resilience evaluation, is presented in [9] where a classification and review of the different approaches proposed in literature are provided. Firstly, the proposed resilience evaluation approaches can be classified as qualitative methods and quantitative methods. Qualitative methods, thorough general picture of the system, provide guidelines for long-term energy policy making. In contrast, quantitative methods are often based on the quantification of system performances. The different methods can be further classified as simulation-based [20], analytical-based [21], and methods based on statistical analysis and historic outage data [22]. According to this classification, RecSIM can be classified as a quantitative-simulation-based approach. In particular, RecSIM takes in input a damage scenario (i.e., the set of electrical grid components in failure), the resource available to face the crisis in terms of crews available, and the functioning status of the supervisory control and data acquisition (SCADA) system and computes, in output, the power grid performance degradation in terms of the number of electrical users disconnected times the minutes of disconnection. As metrics for characterizing, in a posteriori analysis, the resilience of the power grid is proposed in [24] in terms of outage duration, dependency and interdependency relations, and the existence of energy storages, and a mathematical model for their calculation is proposed and implemented with respect to test cases focusing on recent natural disasters hitting major countries. In [25] the authors adopted the definition of resilience provided by the NIAC [26] that considers robustness, redundancy, and rapid recovery as main resilience features and developed a sequential Monte Carlo-based model for assessing the impact of weather on EDN resilience and applied to transmission networks. Their model considers the impact of human response during weather emergencies through the characterization of the delay required for the restoration of damaged components (due to delay in the development of individual situation awareness in the affected control centers) and the delay in the information sharing between the system agents, namely, the transmission system operators (TSOs) and the repair crews. As a test case, the model was applied to the transmission network considering extreme wind events, and simulation results show the resilience of the network in terms of robustness, redundancy, and response measures. Other past works also included the effects of humans [27], and others consider the dependencies [28] on resilience.

Similar to the approach proposed in [24] but considering the performance of EDN grids in complex urban contexts, RecSIM considers, simultaneously, the influence of different key features that might affect the time required for restoring the functionality of EDNs after extreme events, namely, (1) the degree of dependency with other networks providing essential services; (2) the network topology; (3) the number of repair crews available; (4) the number and functionality of SCADA telecontrol devices; and (5) the conditions of the road network and of the traffic that might delay repair activities.

3. Model description

The proposed model aims at providing a model scheme, for the resilience assessment of EDNs, where all the abovementioned influencing factors could be appropriately considered.

Having recognized that resilience mostly starts with a number of activities that are performed during the normal operational mode of the network such as ordinary management of assets, accurate prediction of the events, and subsequent efficacy in performing preparedness actions rather than only with a "last minute" emergency management; the idea was to realize advanced technological tools enabling CI operators to improve the operational procedures during the normal operation mode while ensuring a continuous monitoring of external scenarios to forecast possible perturbing events, accompanied to some ex ante prediction of the expected impact (in terms of both economic losses and reduction of citizen's well-being) of possible emergency scenarios. With this objective in mind, ENEA has designed and realized a decision support system (DSS), called CIPCast, enabling to provide an operational (24/7) forecast and risk analysis for the CI in a specific area [16]. CIPCast includes a map of CI elements which could be hit and disrupted by predicted natural events (flash floods, snow, landslides, flooding) or occurred events (such as earthquakes). CIPCast allows to estimate:

- The physical impacts induced on EDNs following earthquakes [17] and flooding events

- The impact on service functionality associated with the predicted damage of CI elements (in terms of outage duration and geographical extension), also considering possible perturbation cascades toward other networks and services [29, 30]

- The consequences of the predicted outages, according to several metrics accounting for economic losses, reduction of citizen well-being, and impacts on the quality of service

Within CIPCast, the RecSIM simulator represents the basic module for the resilience assessment of the EDN, as better described in Section 4.

This section describes the theoretical model used to represent the topology of a large EDN within RecSIM. **Figure 1** shows the main elements of the proposed model.

EDNs are composed of a number of primary substations (PS). Each PS originates one (or more) medium-voltage (MV) line(s) ending into a further PS. The MV line is cut at a certain stage by a switch which decouples the line into two halves, each one supplied by one of the two overlooking PS. Each line connects a number of secondary substations (SS) that, from the technological point of view, can be of one of the following types: "normal," "remotely telecontrolled," "automated," and "frontier" substations (represented, respectively, as white, gray, orange, and purple nodes in **Figure 1**). The "automated" substations are key elements of the network as they are able to perform automatically the isolation and restoration procedures needed to react to failures happening to their downstream substations. "Frontier" substations can be used to restore a portion of a MV line from another MV line. The configuration of the network switches defines the running configuration of the network. The electrical operator attempts to operate the network in order to maintain as much as possible the grid in a so-called normal configuration which is chosen by the operator as being able to allow the optimal operability of the grid (i.e., a good trade-off between robustness and efficiency, with the lowest possible electrical losses).

During a crisis, the electrical operator can change the configuration of the network by operating the switches along the perturbed lines; the operator brings the network into a "contingency" configuration, in order to restore as fast as possible the electrical service to the final users.

The model considers, furthermore, the dependencies between the electrical distribution grid supervisory control and data acquisition (SCADA) systems and the telecommunication components providing the telecontrol service. As shown in **Figure 2**, the telecontrolled substations use the communication service provided by the telecommunication (tlc) network components (i.e., the base transceiver

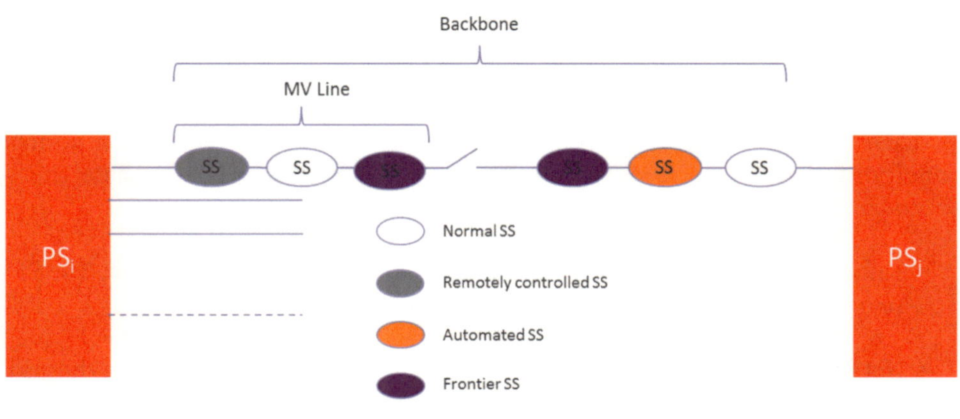

Figure 1.
The elements considered in the electrical distribution grid model.

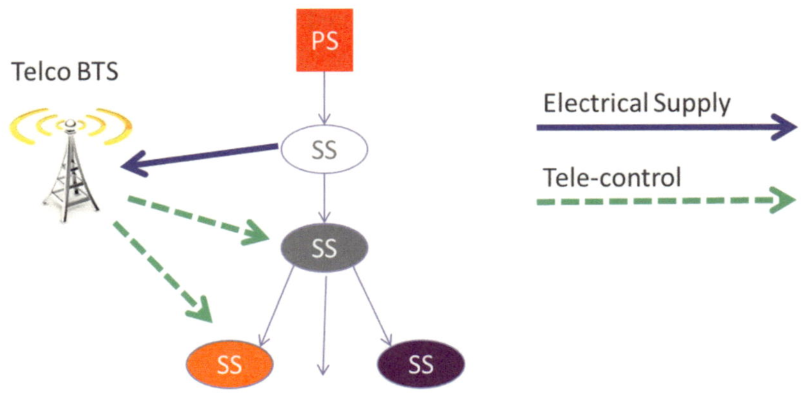

Figure 2.
Electrical distribution grid SCADA system and tlc dependencies.

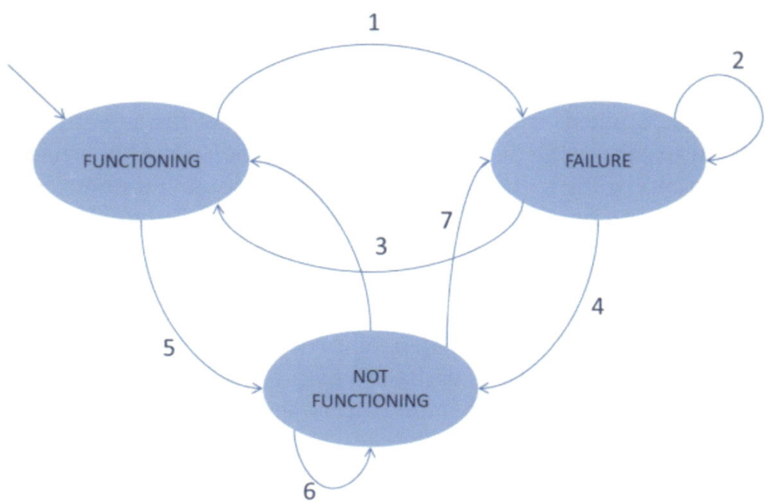

Figure 3.
Secondary substation (SS) finite state model.

station—BTS hereafter—of the telecommunication network). In turn, BTS are supplied by the energy provided by the SS of the EDN, thus configuring a dependency loop (no energy on a specific BTS, no telecontrol functionality of this BTS in favor of other SS of the network). In this work, we suppose that BTS do not have power backup, i.e., we will simulate the worst possible case. This implies that if a certain BTS depends on a certain substation SS that is in a damaged (or disconnected) state, that specific BTS will immediately stop functioning.

Each SS can be modeled as a finite state machine as shown in **Figure 3**. In normal conditions, the SS is in the initial "functioning" state. Starting from this state, the secondary substation (SS) can move into two different states:

- Failure state: when a failure in the SS occurs, transition 1 is activated. The SS remains in this state for the expected failure duration;

- Not functioning state: in case of a contingency, the protection devices of the grid will disconnect a number of secondary substations that will change their state from "functioning" to "not functioning." For example, when a SS moves into the failure state, all SS on the same line move into their not functioning state. A SS remains in this state waiting for restoration. The duration of the restoration can be in the range of few minutes (about 3–5 minutes) if the SS can be telecontrolled or much longer (50–55 minutes to few hours) depending on several factors (e.g., time required by emergency crews to reach the faulted substation and to restore it).

4. Resilience metrics

Let us assume to have an EDN characterized by its topology, with nodes $\boldsymbol{N}$ and links $\boldsymbol{L}$ corresponding to electrical stations and electrical lines, respectively. The function representing the *functioning state* for all the elements of the EDN is referred to as $\boldsymbol{F}$:

$$\boldsymbol{F}(\boldsymbol{N},\boldsymbol{L},\boldsymbol{t}) = \boldsymbol{0}\forall \boldsymbol{t} \tag{1}$$

if all elements $\boldsymbol{N}$ and $\boldsymbol{L}$ are in a *functioning state* and all telecontrol functionalities are active. Let us now introduce a perturbation function $\boldsymbol{P}$ that can change the state of one EDN element from the *functioning state* to one of the other possible states. In such a case

$$\boldsymbol{P}:\boldsymbol{F}(\boldsymbol{N},\boldsymbol{L},\boldsymbol{t}) \rightarrow \boldsymbol{F}'(\boldsymbol{N},\boldsymbol{L},\boldsymbol{t}) \tag{2}$$

where $\boldsymbol{F}'(\boldsymbol{N},\boldsymbol{L},\boldsymbol{t}) > \boldsymbol{0}$ for $\boldsymbol{t} \in [\boldsymbol{0},\boldsymbol{T}]$ and zero elsewhere. For the sake of simplicity, we will apply the perturbation P only to the electrical secondary station (referred to as SS). Time T represents the time when all elements have been repaired and the network comes back to its fully functional state $\boldsymbol{F(N, L, t)} = 0$. A perturbation $\boldsymbol{P}$, in principle, could affect one (or more) electrical station and bring it (or them) from the functioning state to the not functioning or the failure states.

The damage of a SS consequent to the introduction of $\boldsymbol{P}$ produces a sequence of perturbations on the network. These consist in the disconnection of other nodes along the line due to instantaneous opening of protection switches. The damaged nodes are replaced by power generators (PGs) to ensure electrical continuity to the node's customers. The damaged nodes will not be repaired in the time space of the simulation, but their function will be restored through the settlement of PGs. The disconnected nodes, in turn, are reconnected either through a telecontrol operation (if available) or by dispatching technical crews to provide manual reconnection. All such interventions require specific times, which are considered when defining a restoration sequence of interventions. The impact of the perturbation $\boldsymbol{P}$ on the EDN is measured using a key performance indicator (KPI) that is currently used by the Italian Energy Authority to estimate the level of service continuity of an EDN. Such KPI is expressed as the number of disconnected customers $\boldsymbol{ni}$ of the *i-th* EDN node times the duration τ_i of its disconnection. Such a value is expressed in terms of *kilominutes* (i.e., 10^3 minutes). Thus, if the damage of the *i-th* SS of the network will result in the disconnection of $\boldsymbol{m}$ SS, each one for a time τ_j $(\boldsymbol{j} = \boldsymbol{1},\boldsymbol{m})$, the overall KPI outage metrics will be measured in terms of Γ_i that is defined as follows:

$$\boldsymbol{\Gamma}_i = \sum_{j=1}^{m} \boldsymbol{n}_j \boldsymbol{\tau}_j \tag{3}$$

For a given perturbation $\boldsymbol{P}$, the integral over the simulated time span of Eq. (3) represents the perturbed functional state of the grid defined in Eq. (2):

$$\int_0^T \boldsymbol{F}'(\boldsymbol{N},\boldsymbol{L},\boldsymbol{t})\boldsymbol{dt} = \boldsymbol{\Gamma}_i \tag{4}$$

Γ_i, thus, represents the ***impact*** that the damage of an EDN element (the *i-th* node) can produce, by using an official KPI as a metric. The larger the value of Γ_i, the weaker the capability of the network to withstand the perturbation in terms of impacts produced on the EDN customers. In general the value of Γ_i depends on different factors (described in detail in Section 4) ranging from the topology of the network and the employed technologies to the efficiency of operator restoration procedures; therefore, it would not be inappropriate to correlate the value of Γ_i with the inverse of the resilience concept ***R***. In other terms

$$\boldsymbol{R}^{-1} \propto \boldsymbol{\Gamma}_i \tag{5}$$

We can generalize the concept by checking the EDN behavior versus all possible perturbations. The overall operational network resilience will be thus associated with the inverse of the value of the integral of the distribution function of all the Γ_i values ($D(\Gamma)$) resulting from the failure of each one of the *N* nodes of the EDN (normalized with respect to the total number of nodes *N*):

$$\boldsymbol{R} \propto \frac{1}{<\boldsymbol{\Gamma}>} = \frac{\int D(\boldsymbol{\Gamma})d\boldsymbol{\Gamma}}{\int \boldsymbol{\Gamma} D(\boldsymbol{\Gamma})d\boldsymbol{\Gamma}} \tag{6}$$

The higher the impact, the lower is the resulting operational resilience of the EDN network.

5. The simulation scheme and the reconfiguration simulator (RecSIM)

The RecSIM represents the basic module of the proposed framework for the resilience assessment of the EDN. RecSIM enables to carry on a "crisis game" consisting in the estimation of all Γ values resulting from the application of different perturbations. The simulator allows configuring different parameters allowing, in turn, the simulation of different electrical operational conditions (e.g., SCADA system not available, traffic jams, etc.) and the analysis of how the resilience indicator varies in these different operation conditions.

Figure 4 shows the input of the RecSIM and its output (i.e., the consequence of a perturbation in terms of Γ_i). RecSIM inputs are:

- *Network topology*—expressed as the EDN graph and the perturbation *P* represented by the SS brought in the damaged state.
- *SCADA system*—expressed in terms of the set $\boldsymbol{\Omega}$ of SS that can be remotely telecontrolled.
- *Efficiency of SCADA system*—expressed in terms of the functioning status of the BTS b_i providing communication service to the EDN and in terms of tlc_t the time needed to perform a remote operator action (using the EDN SCADA functionalities).
- *Efficiency of restoration procedures*—expressed in terms of the time needed by an emergency crew to reach a damaged SS ($\boldsymbol{tr}_t$), to perform a manual reconnection action ($\boldsymbol{m}_t$), and to set in place a PG to feed the users of the damaged SS (or of other SS which will result to be isolated, thus needing a PG as they were damaged) (PG_t). The input time values represent "mean" values as they have been provided by the electrical operator as resulting from its standard

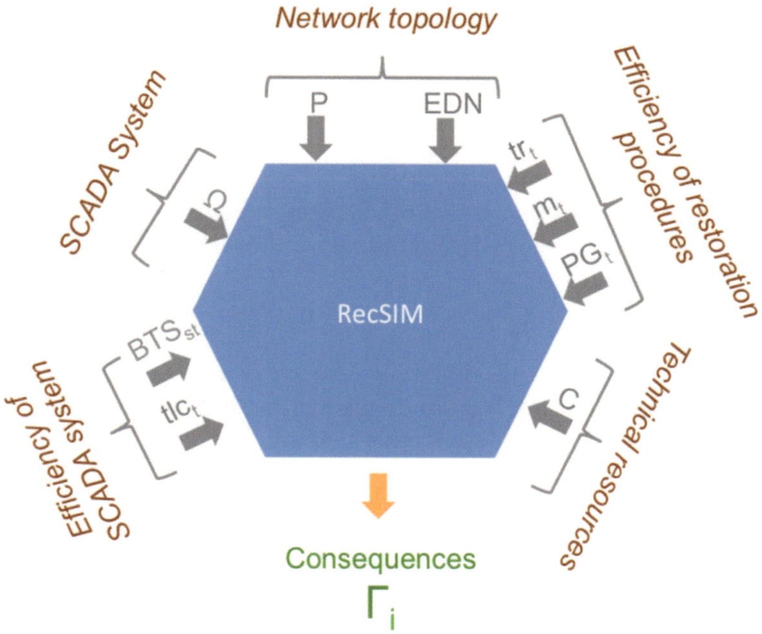

Figure 4.
The RecSIM input and output components.

operation times. RecSIM performs simulations by using these values as mean values of a flat distribution from which time values to be used in the simulation are randomly extracted from the flat distribution.

- *Technical resources*—expressed in terms of the number C of technical crews available in the field. The number of available PGs is assumed to constitute an unlimited resource. Further development of the algorithm will consider the finiteness of available PGs.

The output of RecSIM is represented by the value of the impact of the damage scenario (caused by the perturbation $\boldsymbol{P}$ and by its cascading effects) on the EDN, considering all the restoration actions performed (in series or in parallel, if several technical crews were simultaneously available): the substitution with a PG of a damaged node and, whenever the case, of an isolated node; the manual reconnection of disconnected nodes by the available technical crews; and the automatic reconnections made through remote telecontrol operations. These actions are needed to restore the EDN and to bring it back to its normal operating condition. Upon these actions, all users are supposed to be reconnected to the grid. As previously said, damaged SS are just substituted by a PG, and, at the end of the simulation, they are still in the damaged state although their function is guaranteed by the PG. The impact of the perturbation $\boldsymbol{P}$ on the EDN is thus computed using Eq. (3).

6. Simulation, analysis, and discussion of the results

In a previous work [19], the EDN of the metropolitan Rome, area (Italy), was deeply investigated by extensive calculations enabling to estimate its resilience score, according to the definition reported in Section 3. RecSIM has been used to study the behavior of the whole EDN of the metropoli-Rome EDN that is a large EDN grid composed of 139 PS and 14938 SS distributed along 1607 MV lines.

The power grid has 6348 telecontrolled SS (i.e., 42% of the total SS) and 1012 automated SS (i.e., 6% of the total SS). Considering the MV lines, the power grid has 1447 MV lines that contain at least 1 telecontrolled SS (90%) and 510 MV lines containing at least 1 automatized SS (31%). The considered power grid is set in the so-called normal configuration that is:

1. A specific topology of the network (consisting of a given number of SS, with a given fraction of telecontrolled, automatic, and frontier SS)

2. With the switches along the medium-voltage lines located in specific points,

3. The telecontrolling BTS providing services to a certain extent (in our simulation we consider a default fraction of unavailable BTS leading telecontrolling functionality unavailable—apparently a "physiological" conditions of dependent networks)

4. A given number of technical crews available in the field for the manual recovery operations

5. Standard times for the solution of the different actions to be performed for SS restoration

When referred to "normal configuration," we will refer to equal (1)–(5) conditions. The reported simulation has the character of a "stress test." Two different stress schemes have been adopted: the *unbiased* perturbation scheme and the *heuristic* scheme.

In the *unbiased* perturbation scheme, each electrical substation (SS_i i = 1, ***N***), one at a time, has been set in the damage state and the resulting impact of electrical crisis estimated in terms of the Γ_i defined in Eq. (3). **Figure 5** reports the distribution function $D^1(\Gamma)$ for all resulting Γ_i. This simulation will be referred to as the "(***N***− 1) analysis," as it involved the set in the damage state of a single SS (at a time).

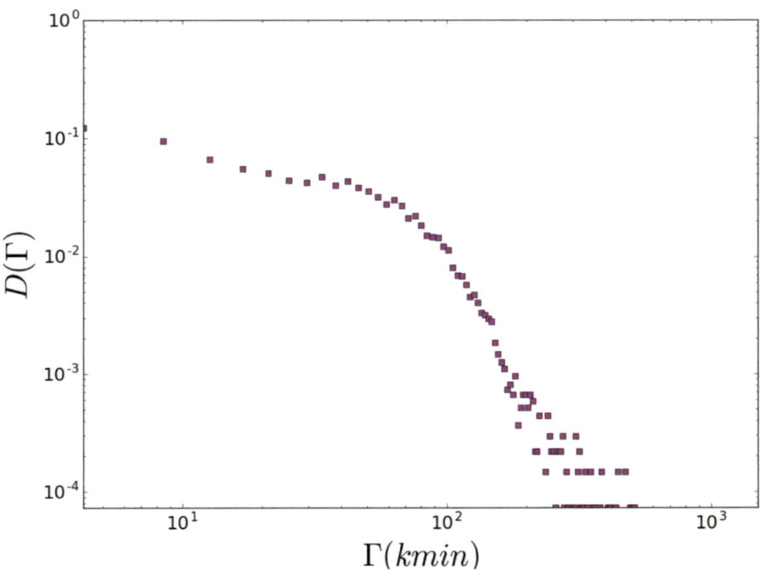

Figure 5.
$D^{(1)}(\Gamma)$ distribution of the resulting impacts as functions of the impacts, for a N − 1 analysis, i.e., in the event of one node damaged.

The same stress test could be repeated by setting in the damage state two, three, or more SS simultaneously, in a way intended to generate crisis situation of higher impact (although with a lower probability of occurrence. Each case will produce a distribution function of the Γ values such as $D^{(2)}(\Gamma)$, $D^{(3)}(\Gamma)$, etc.

The impact distribution $D^{(i)}(\Gamma)$ functions do provide the generating function for the resilience score, which has been associated to its normalized integral Eq. (6): the larger the integral of the distribution, the lower the resilience. In fact, for an infinitely resilient network, each damage should correspond to the lowest possible (or vanishing) impact in the terms expressed by Eq. (3). The overall system resilience could be estimated as a series of terms each one representing the contribution toward resilience for different (and progressively large) perturbations:

$$\boldsymbol{R} = \boldsymbol{a}^{(1)}\boldsymbol{R}^{(1)} + \boldsymbol{a}^{(2)}\boldsymbol{R}^{(2)} + \boldsymbol{a}^{(3)}\boldsymbol{R}^{(3)} + \dots \tag{7}$$

While the terms $\boldsymbol{R}^{(i)}$ will be achieved by applying Eq. (6) to the different $D^{(i)}(\Gamma)$, the terms $\boldsymbol{a}^{(i)}$ can be related to the probability of the event; this would produce a series of progressively smaller terms which will reduce the impact of the high order contributions to the total value. The first terms related to $(\boldsymbol{N} - 1)$ and $(\boldsymbol{N} - 2)$ events would thus dominate the series in Eq. (7) which will provide an *unbiased* estimate of the global resilience of the network when perturbations are imposed following an exhaustive scheme rather than a heuristic method.

A different perturbation scheme (the *heuristic* one) has been also applied to provide a further possible perturbation scheme aiming to realize a resilience assessment which, in such case, will be measured by estimating Eq. (6): this will be done through the use of a distribution function $D^{(h)}(\Gamma)$ resulting from the application of the *heuristic* perturbation scheme. The *heuristic* scheme has been thus designed and applied to compare the resulting resilience score $\boldsymbol{R}^{(h)}$ with that obtained through the use of the *unbiased* perturbation scheme.

Instead of producing systematic damages (as in the *unbiased* scheme), we have produced "educated" damage scenarios where SS have been set in the damaged state as a function of their effective rate of faults (as declared by the electrical operator). The heuristic perturbation scheme is thus carried out in the following way. Let us assume to know the *rate of faults per day* $\boldsymbol{\rho}\mathbf{I}$ of each SS of the network, expressed in terms of the average number of times that the SS have been recorded to be out of order. Statistics have been collected along several years and the number of observed faults normalized over the number of days of observation. The ρ_I value could be thus assimilated to the daily probability that the specific substation goes in a damaged state. The cause of SS fault could be different: the SS could be hit by some external event (i.e., natural hazard and/or its consequences) or by some internal event (i.e., the disruption of some component). The statistical fault rate per component does not distinguish between the origins of fault; we will thus consider this fault rate as an "intrinsic" property of the EDN element.

The *heuristic* perturbation scheme has been thus applied to the network "normal configuration" by simulating $\boldsymbol{M}$ working days: in each day of operations, the damaged state of each SS has been sampled (as in a Monte Carlo scheme) by extracting a random number r_i $(r_i = [0, 1])$ and by comparing it with the $\boldsymbol{\rho}_i$ value: if $r_i < \rho_i$, the i-th SS is put in the damaged state where it remains unperturbed elsewhere. The SS set in the damage state have been put simultaneously in the damaged state, in order to simulate the worst-case scenario. This procedure is repeated $\boldsymbol{n}$ times to scan each SS and then repeated $\boldsymbol{M}$ times to simulate different working days.

This procedure generates very few damages, as the rate of faults of the substations is usually particularly small. However, it generates cases where one (or even more than one) substation will result in a damaged state. This procedure thus allows to sample

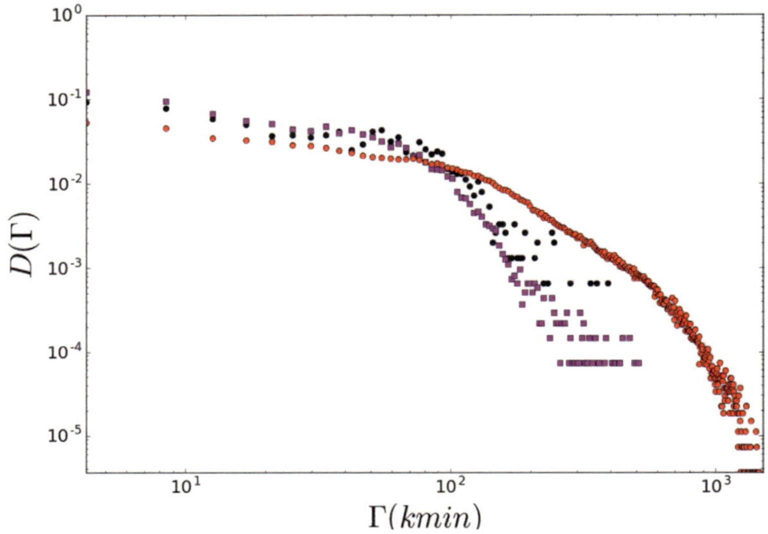

Figure 6.
Comparison of the $D(\Gamma)$ distribution values for the $N-1$ and $N-2$ unbiased scheme simulation resulting from the simulation via the heuristic scheme (red = unbiased ($N-2$), purple = unbiased ($N-1$), and black = heuristic scheme).

(among the manifold of possible damaged network states) those states where one or more SS are simultaneously damaged, in agreement with the rate of faults of the different stations. Over n^h = 1515 damaged configurations were obtained with the Monte Carlo sampling, of which 1163 were constituted by a single damaged SS; 296 with 2 damaged SS; 49 with 3 damaged SS; 5 with 4 damaged SS; and 2 with 5 damaged SS.

Figure 6 summarizes all the results obtained thanks to the simulations by using the ($N-1$) and the ($N-2$) *unbiased* scheme and the *heuristic* perturbation scheme. In all simulations (both for the *unbiased* and for the *heuristic* schemes), the same number of technical crews C available for the service restoration has been assumed ($C = 2$). The three curves, however, derive from simulations scheme which have produced different amounts of crisis scenarios whose impacts have been measured through Eq. (3). In fact, for the *unbiased* ($N-1$) simulation, a number of crisis scenarios n equal to the number of nodes N have been produced ($n^{(N-1)} = N = 13{,}618$). In the case of the unbiased ($N-2$) simulation, a number of crisis scenarios $n^{(N-2)} = 271{,}581$ have been produced (this number corresponds to the total number of double faults occurring along the same medium tension line).

For the heuristic perturbation scenario, the number of cases was, in turn, n_h = 1515 as previously stated. The most relevant feature of the three distributions must be observed in the impact dimension. The perturbations produced by using an unbiased ($N-2$) scheme produce very large effects, as they tend to involve a large number of SS, which impose a sequence of interventions (with the provided number of technical crews k available, not all SS could be simultaneously repaired).

The estimate of the corresponding $R^{(1)}$, $R^{(2)}$, and $R^{(h)}$ [through the use of Eq. (6)] appropriately renormalized all the distributions. Application of Eq. (6) to the three different distribution functions provides the following values: $R^{(1)} = 2.17 \times 10^{-2}$, $R^{(2)} = 7.60 \times 10^{-3}$, and $R^{(h)} = 1.78 \times 10^{-2}$.

It is interesting, in turn, to notice that crises produced by the *heuristic* scheme (i.e., involving SS which have shown a large propensity to fault), although in some cases involving more than a SS produces impacts which, even in the largest cases, are of the same dimension of those produced by worst cases in the ($N-1$) unbiased simulation. This is probably due to the fact that more vulnerable SS are located along lines, which do not produce relevant outages in case of fault (either for the presence of a few not remotely controlled SS and/or for the presence of a not too large number of connected customers).

Different scores are the results of the different adopted simulation schemes. Rather than the absolute resilience score, what should be estimated which might have a technological meaning are “resilience score variations”: when the same network (and/or its management properties) is modified, the same simulation scheme can be adopted and the resilience score measured again. The difference of the resilience score (before and after the modifications) will provide an indication on if modifications have (or have not) produced benefits to the overall network resilience.

7. Conclusions

The work presented in this chapter that built a great amount of work done on the same topic [18, 19] presents the RecSIM system and its relevant capabilities to represent and simulate real urban system and in particular problems related to the reconfiguration of electric distribution systems following faults. In particular two major achievements are highlighted, one related to the possibility to account for a number of issues, which have not been appropriately considered in the resilience assessment process in the existing literature, and the second concerning the viability of implementing RecSIM (and its scalability) to large, real EDN. In particular reference has been made in the paper to the case study of Rome city that has a quite large distribution network containing more than 13,500 electrical substations.

As for the general achievements in the area of the models for estimating resilience of EDN, a novel, computable scheme has been identified, on which the RecSIM engine, described in the paper, is based on. The RecSIM model considers different factors encompassing all the phases of risk management, including the technological properties of the network, the fault management procedures, and the network interdependency with the telecontrol network. In many cases of previous works on the same topic (recalled in Section 2), the resilience estimates have been done by using models which considered just the electrical response of the network, thus disregarding the topological and technological features of the network, as well as the management skills and procedures, and the external and environmental constraints. The EDN management model behind the RecSIM tool, in turn, is able to reconstruct the impact of a crisis by considering all the abovementioned factors (recalled in Section 4) which play a critical role in determining the overall systemic resilience of the EDN. Moreover, the possibility of relating the resilience to the distribution of impacts generated by a range of possible perturbations, described in this chapter, provides a further improvement to the prosed approach. Many different perturbation schemes could be therefore investigated, and a resilience score, more suitable for to the user’s requirements, can be therefore assessed. Last but not the least, this scheme could also be prone to be modified by varying the outage impact metrics. Whereas in this work the outage impact Γ was assessed in terms of the KPI adopted by the Italian regulatory agency [Eq. (3)], it can be expressed by considering further metrics, able to account, for instance, the economic losses or the level of wealth reduction caused to the citizens [19].

As from the analysis of the data resulting from the case study analyzed, i.e., the Rome city EDN, the profile of the impact distribution functions resulting from the different simulations made on the basis of the *unbiased* and the *heuristic* schemes has revealed two main results.

Firstly, the *unbiased* ($N-2$) scheme provides the worst-case scenario. The simultaneous damage of two SS residing along the same medium-voltage line, produces (as expected) impacts of a significant severity since several other SS are involved.

In this case, the model would be able to help the detection of the most impacting causes and to validate the possible improvements which could be introduced by acting on specific issues (i.e., by increasing the quantity of telecontrolled SS along the lines and/or by increasing the number of technical crews available and/or by improving the telecontrol strategy). This information would be particularly useful for electrical operators for the planning of new activities for enhancing resilience.

Secondly, the *heuristic* scheme, where SS are damaged according to their effective rate of fault (as measured and reported by the electrical operator), provides a resilience score which is slightly lower than the one resulting from the ($N-1$) unbiased scheme. As previously discussed, this could be the result of the correct management of the operators which has "segregated" more vulnerable assets along the lines whose disruptions cause less relevant impacts on services. The RecSIM tool, in this respect, could be useful for assessing which should be the correct way for further improving this score by selecting the substations (among those which have produced the crisis scenarios accounted for in the simulations) whose robustness improvement could further reduce the impact and thus increase the resilience score. Moreover, the tool can be used within more general framework as, for example, the emergency management support tool CIPCast-ES [16] which allows to explore a realistic earthquake event occurring in an urban area by predicting disruptions on buildings and critical infrastructure and by designing a reliable scenario, accounting for road obstruction due to building collapse, to be used to design an efficient contingency plan.

In conclusion, the RecSIM model, being able to gather into a unique scheme several EDN features, can provide a reliable tool for the analysis of large and complex infrastructures. This is going to be exploited in Italy through the establishment of a competence center for risk analysis and forecast of critical infrastructure called EISAC.it (*European Infrastructure Simulation and Analysis Centre Italian node* [31]) which will deliver competences and services to support operators and public authorities committed to the protection and the emergency management of critical infrastructure.

Acknowledgements

This work has been triggered by the project CIPRNet (funded by EU under the FP7 program through GA no. 312450) and the Italian (MIUR) funded project RoMA (SCN_00064): all project partners are acknowledged for their support in the RecSIM elaboration. Areti SpA, the electrical distribution company of the city of Rome, is also acknowledged for the available network's data, in particular ING. Silvio Alessandroni is acknowledged for his constant support and for being the source of illuminating suggestions and comments.

List of acronyms

EDN	electrical distribution network
CI	critical infrastructures
TSO	transmission system operator
DSS	decision support system
PS	primary substations
MV	medium-voltage line(s)
SS	secondary substations
SCADA	supervisory control and data acquisition
tlc	telecommunication network components
BTS	base transceiver station

N	nodes
L	links
F	functioning state function
P	perturbation function
t	time
T	time when the functioning of the system is re-established
PG	power generator
KPI	key performance indicator
ni	number of disconnected customers
$\boldsymbol{\tau}_i$	disconnection duration
$\boldsymbol{\Gamma}_i$	impact
R	resilience
$\mathbf{D}(\boldsymbol{\Gamma})$	distribution function of all the values
$\boldsymbol{\Omega}$	set of secondary substations
$\boldsymbol{tr_t}$	time required to reach a damaged link or node
$\boldsymbol{m_t}$	time required to perform a manual reconnection action
PG_t	time required to install power generator backups
C	number C of technical crews available in the field
$\mathbf{D}^1(\boldsymbol{\Gamma})$	distribution function for the resulting impacts
(N-1)	analysis performed in the hypothesis of one nonfunctioning node
a(i)	factor related to the probability of a crisis event
(N-2)	analysis performed in the hypothesis of two nonfunctioning nodes
$D(h)(\Gamma)$	distribution function for the resulting impacts for an heuristic perturbation scheme
ri	random number
M	M working days
n	number of crisis scenarios

Author details

Alberto Tofani, Gregorio D'Agostino, Antonio Di Pietro, Sonia Giovinazzi, Luigi La Porta, Giacomo Parmendola, Maurizio Pollino and Vittorio Rosato*
Laboratory for the Analysis and Protection of Critical Infrastructure, ENEA Casaccia Research Centre, EISAC.it, Rome, Italy

*Address all correspondence to: vittorio.rosato@enea.it

References

[1] Karagiannis GM, Turksezer ZI, Alfieri L, Feyen L, Krausmann E. Climate change and critical infrastructure-floods. In: EUR-Scientific and Technical Research Reports. Luxembourg: Publications Office of the European Union. 2017

[2] Giovinazzi S, Wilson T, Davis C, Bristow D, Gallagher M, Schofield A, et al. Lifelines performance and management following the 22 February 2011 Christchurch earthquake, New Zealand: Highlights of resilience. Bulletin of the New Zealand Society for Earthquake Engineering. 2011;**44**:402-417

[3] Kwasinski A, Eidinger J, Tang A, Tudo-Bornarel C. Performance of electric power systems in the 2010-2011 Christchurch, New Zealand, earthquake sequence. Earthquake Spectra. 2014;**30**(1)

[4] Kongar I, Giovinazzi S, Rossetto T. Seismic performance of buried electrical cables: Evidence-based repair rates and fragility functions. Bulletin of Earthquake Engineering. 2017;**15**:3151

[5] Calcara L, Di Pietro A, Giovinazzi S, Pollino M, Pompili M. Towards the resilience assessment of electric distribution system to earthquakes and adverse meteorological conditions. In: 10th AEIT International Annual Conference, AEIT. 2018

[6] Bruneau M, Chang SE, Eguchi RT, Lee GC, O'Rourke TD, Reinhorn AM, et al. A framework to quantitatively assess and enhance the seismic resilience of communities. Earthquake Spectra. 2003;**19**:733-752

[7] Haimes YY. On the definition of resilience in systems. Risk Analysis. 2009;**29**(4):498-501

[8] Eidinger J, Tang A. Resiliency of Power Grids after Earthquakes, Applied Technology Council; 2016

[9] Bie Z, Lin Y, Li G, Li F. Battling the extreme: A study on the power system resilience. Proceedings of the IEEE. 2017;**105**:1253-1266

[10] SMR EU Project. Available from: http://smr-project.eu/home/

[11] Improver. Improver EU Project. 2017. Available from: http://improverproject.eu

[12] Darwin EU Project. Available from: https://h2020darwin.eu

[13] Resilens EU Project. Available from: http://resilens.eu

[14] Smart Resilience EU Project. Available from: http://www.smartresilience.eu-vri.eu

[15] Resolute EU Project. Available from: http://www.resolute-eu.org

[16] Di Pietro A, Lavalle L, La Porta L, Pollino M, Tofani A, Rosato V. Design of DSS for supporting preparedness to and management of anomalous situations in complex scenarios. In: Setola R, Rosato V, Kyriakides E, Rome E, editors. Managing the Complexity of Critical Infrastructures: A Modelling and Simulation Approach. Cham: Springer International Publishing; 2016. pp. 195-232

[17] Giovinazzi S, Pollino M, Kongar I, Rossetto T, Caiaffa E, Pietro AD, et al. Towards a decision support tool for assessing, managing and mitigating seismic risk of electric power networks. In: Lecture Notes in Computer Science. 2017

[18] Di Pietro A, Porta LL, Lavalle L, Pollino M, Rosato V, Tofani A. Simulation of cascading outages in (inter)-dependent services and estimate of their societal consequences. In: Havarneanu G, Setola R, Nassopoulos H, Wolthusen S, editors. Critical Information Infrastructures Security. CRITIS 2016. 2017

[19] Tofani A, D'Agostino G, Di Pietro A, Onori G, Pollino M, Alessandroni S, et al. Operational resilience metrics for a complex electrical network. In: Critis 2017 Lecture Notes in Computer Science (LNCS) Series. 2018

[20] Watson JP, Guttromson R, Silva-Monroy C, Jeffers R, Jones K, Ellison J, et al. Conceptual framework for developing resilience metrics for the electricity oil and gas sectors in the United States. In: Sandia Report Sand 2014-18019. 2014

[21] Shinozuka M, Chang SE, Cheng T, Feng M. Resilience of integrated power and water systems. In: Proc. Seismic Eval. Retrofit Lifeline Syst. 2003. pp. 65-86

[22] Chanda S, Srivastava AK. Defining and enabling resiliency of electric distribution systems with multiple microgrids. IEEE Transactions on Smart Grid. 2016;**7**(6):2859-2868

[23] Willis HH, Loa K. Measuring the Resilience of Energy Distribution Systems. Tech. Rep. 2015. Available from: https://www.rand.org/pubs/research_reports/RR883.html [Accessed: 01 March 2018]

[24] Kwasinski A. Quantitative model and metrics of electrical grids: Resilience evaluated at a power distribution level. Energies. 2016;**9**(12):93. DOI: 10.3390/en9020093

[25] Panteli M, Mancarella P. Modelling and evaluating the resilience of critical electrical power infrastructure to extreme weather events. IEEE Systems Journal. 2017;**11**(3):1733-1742

[26] National Infrastructure Advisory Council (NIAC). A Framework for Establishing Critical Infrastructure Resilience Goals; USA; 2010

[27] Zio E. Reliability engineering: Old problems and new challenges. Reliability Engineering and System Safety. 2009;**94**:125-141

[28] Reed DA, Kapur KC, Christie RD. Methodology for assessing the resilience of networked infrastructure. IEEE Systems Journal. 2009;**3**:174-180

[29] D'Agostino G, Di Pietro A, Giovinazzi S, La Porta L, Pollino M, Rosato V, et al. Earthquake simulation on urban areas: Improving contingency plans by damage assessment. In: Luiijf E, Žutautaitė I, Hämmerli B, editors. Critical Information Infrastructures Security. CRITIS 2018. Lecture Notes in Computer Science, Vol. 11260. Cham: Springer; 2019

[30] Di Pietro A, Porta LL, Lavalle L, Pollino M, Rosato V, Tofani A. Simulation of cascading outages in (inter)-dependent services and estimate of their societal consequences. In: Havarneanu G, Setola R, Nassopoulos H, Wolthusen S, editors. Critical Information Infrastructures Security. CRITIS 2016. Lecture Notes in Computer Science, Vol. 10242. Cham: Springer; 2017

[31] EISAC. Available from: www.eisac.it

Chapter 4

Measuring Infrastructure Skills Productivity

Rex Asibuodu Ugulu, Stephen Allen and Andrew Arewa

Abstract

This chapter presents the concept and meaning of construction productivity and the techniques used in measuring workforce skills productivity in the construction industry. There are three major methods highlighted for monitoring and measuring productivity in the construction industry. The first relates to visual recording that requires taught watchers to be available on location to monitor and record work by specialists. The second is the physical recording technique which involves a direct surveillance technique that includes a qualified observer observing the site for the full duration of the working day operation using the work study method or work measurement. This method was typically illustrated with the application of regression model and learning curve theory to improve labour productivity. The third measuring technique discussed is the use of questionnaires and interview survey which involves information gathering through an interview with supervisors and workers working in the construction industry. Finally, the chapter discussed how infrastructure productivity can be improved through effective supervision, material management and supply chain management, project front-end planning (loading) and work face planning, training and certification of workforce and labour management and relations.

Keywords: construction, measurement, productivity, method study, skills, work measurement, work study

1. Introduction

Productivity in its broader sense is the association between a given output and the resources used to produce it; this relationship represents the global productivity factors, expressed as a fraction, of which the numerator is a measure of output and the denominator a measure of all factors combined. The combination of the factors of production (capital, labour, raw materials and power) is always heterogeneous, and values must be introduced to find an expression for it [1]. It is easier to relate output to a single factor than to a combination of factors. Labour is the factor considered in most cases, but it may be capital or a given raw material, for example.

Productivity is defined by the *Concise Oxford English Dictionary* "as the effective power of production and the proportion which goods are produced". In this definition, it can be deliberated that the main resource in productivity is the workforce.

Construction operatives are the most productive resource; therefore, productivity in the construction industry is dependent primarily on human performance and effort [2]. Construction labour productivity is an important productivity index due to the numerous human labour required to carry out and complete a task. Researchers in the field of economists will agree on the significant of productivity to an industry, an individual enterprise or the national economy; however, there has not been any agreement on the actual measurement techniques of productivity.

The Bureau of Labor Statistics (BLS) updated the productivity measures for four infrastructure construction industries in 2017. These industries were only recently published due to difficulties in measuring worked hours and output [3]. The US construction sector in 2017 contributed a large portion to the economy with 4.3% of GDP and 6.1% of all industry employment which was attributable to the sector. Recently, the BLS noted that the infrastructure industries comprise about 10.5% of the entire industry sector's employment.

1.1 Productivity concepts

There have been several construction labour productivity (CLP) definitions that point towards the various perception of the productivity in the construction industry. The European Association of National Productivity Centres [4] defined productivity "as how resourcefully and efficiently goods and services are being formed, efficiently in this setting be doing things right or utilizing resources to achieve anticipated results". Productivity is also defined by the American Association of Cost Engineers [5] "as a labour efficiency of relative measure, either bad or good when related to a recognised base or norm." "The comparative nature of construction productivity generate most challenges in tracing it as a complete significant over time, it is feasible to gather information on the arrangement of the recognised base, or yardstick" [6]. However, the professionals in the construction industry and project managers defined labour productivity as a proportion of expected over actual productivity; this is mathematically expressed as

$$\text{Performance Ratio (PR)} = \frac{\text{Actual Productivity}}{\text{Expected Productivity}} \quad (1)$$

When a PR is higher than the unity, it means that the daily-based quantities require more work hours than the normal average day baseline; also, it implied that the productivity of that day was worse than the baseline productivity [6]. The importance of this method is that progress is based on the constructed work and not the hour work utilised and productivity can be measured despite the type of work done. The Bureau of Labor Statistics [7] well defined labour productivity as productivity per real worked hours, and these hours refer to the really worked hours, and this quantity discounts holidays, trip and sick leave; nevertheless, it comprises salaried and voluntary overtime.

Productivity measures how capitals are hired efficiently; it is defined "as the quantity of a specific amount of input per unit of labour to a specific measure of output and usually measured as total output divided by the numbers of units of labour employed to produce that output" [8]. Labour productivity in the construction industry can be described as the man-hour output per day, which is often reduced by poor provision of inadequate tools and equipment; delayed, unclear or inadequate instructions; unbalanced work gangs; wrong working method; inadequate incentives; and non-delegation of authority from senior supervisors to lower supervisors.

Productivity definitions seem to be dependent on the context and the researcher's opinion on the study. Previous studies on engineering, technology and economics classified productivity into three broad industry groups; these various groups defined productivity from their various viewpoints [9].

Definitions of productivity are aimed at explaining the meaning of the term, while measured descriptions in the previous case are used as a base of measurement; the main objective is to increase and productivity [10].

From the above background, productivity can be defined as the volume correlation relating to work hour utilisation and production. It is also the correlation relating to amounts of input hours used by an organisation to produce output:

$$\text{Labour productivity} = \frac{\text{output produced}}{\text{input work - hour}} \tag{2}$$

From Eq. 2, output can be described as any process of the outcome, either a product or service, whereas input work hour consists of the hours utilised in a process. However, the relationship between a variety of output components and work hour is very complex. Measurement of construction skills productivity presents practitioners and academics with a variety of challenges.

Several challenges have been recognised by diverse researchers for declining labour productivity under the classifications of incomplete documentation, design changes, supply chain fragmentation and inefficient project management as the most important drivers affecting productivity in the mid-rise private development in Australia [11]. Ugulu and Allen [12] carried out an experimental examination on the importance of on-site craft gangs' learning productivity and found an average improvement of learning rate of 94.21% gains as illustrated in **Table 3**. The study additionally discovered that on-site learning is a huge factor influencing productivity of construction craft gangs. In a related study, Tanko et al. [13] developed a framework for value management (VM) implementation, exploring the current construction practice factors that will improve construction productivity, the study found that people, environment, government and information are important factors that can be used by construction stakeholders to improve construction productivity practices in the Nigerian construction industry.

An investigation on productivity of panelised and long span timber construction using time lapse photography on five active case study construction sites to measure installation productivity was conducted by Forsythe, Brisland [14], using net crane time as the basis for measuring productivity, being the time dedicated purely to crane cycles involved directly in installing prefabricated timber panels; 521 cycles were measured relating to the installation of 5592 m^2 of panels. The study found that panelised prefabricated timber construction offers a fast and productive site installation process [14].

2. Measuring construction productivity

Labour productivity measurement in the construction industry is a difficult task; in most cases labour productivity is usually taken to mean productivity. This stems from the fact that projects in the construction industry are labour-intensive using basic equipment [15]. Productivity described the quantification of how well operatives use available resources to produce outputs from inputs [15]. The American Association of Cost Engineers [5] described productivity as when associated with standard or to a recognised base as the relative measure of labour efficiency is either good or bad.

The BLS [3] opined that the increase in labour productivity during the 2000–2005 period was primarily driven by a large rise in output. Beginning in 2005, output fell through 2009 at a speedy rate than the worked hours, leading to a sharp reduction in productivity of labour during that period. The productivity of labour

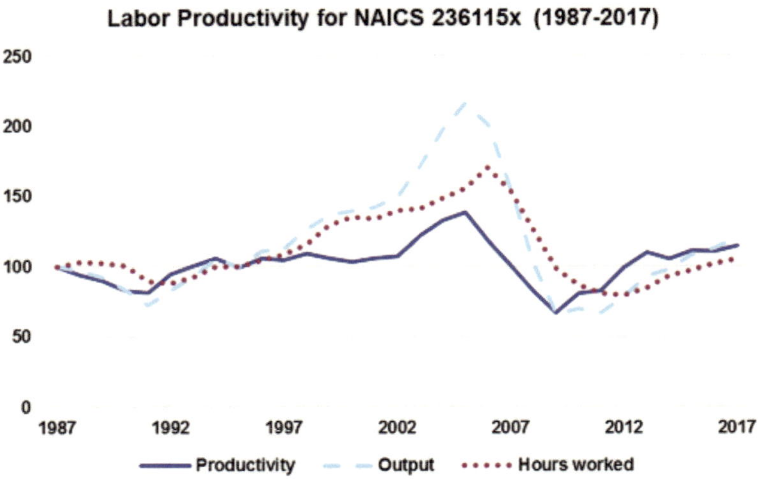

Figure 1.
Productivity of labour for single-family residential infrastructure [1].

increased since 2009; however, in 2017 it is far below its 2005 topmost as presented in **Figure 1**. productivity of labour for single-family residential infrastructure (NAICS 236115x). The report further stated that infrastructure industry experienced a decrease in productivity output and a slight fall in worked hours, resulting to decreasing productivity of labour until 2010. In 2010, the productivity output returns to normal until 2017, when output slightly declined [3].

Construction skills productivity measurement is complex and challenging when comparing productivity in the construction industry between nations [16]. Productivity indices are utilised to adjust input and output information with the goal that efficiency measurement can be differentiated after some time and among industry divisions, and countries found that these lists vary among nations [17]. For instance, output in the construction industry is complicated when compared and quantified within a nation due to complexity in comparing single-family households to transportations, colleges to bridges or clothes centres to workplace buildings.

The National Research Council [18] noted that worldwide productivity evaluations have similar encounters, such as the heterogeneity of inputs and outputs. The loss of valued information on the nature of heterogeneous building construction output is as a result of combining data to an organisational level [17]. In addition, [18] noted that "an industry productivity analyst varies on whether productivity in the construction industry is decreasing or improving." Productivity has been declining for over 30 years. However, some construction projects and construction tasks studies document investigated improved in their productivity in the US industry productivity analysis [18].

Contractor's labour performance measure is degraded by variability in building processes when the presence of variation with process of construction time can no longer be determined [19]. Construction variability inflates the construction duration and decreases the processes and the ability of the production network [20]. Rework flow in construction project is a significant cause of variability that causes procedures to be unpredictable [6]. However, in building projects, rework can be arising due to errors in construction exposed through official stage examinations or casual on-site observations. Additionally, another kind of rework is client-related rework, which is instigated by variations in project design, plan and scope by the client [21].

Construction on-site is environmental dynamic and subjected to a high level of unpredictability and outside volatility that can be caused by constraints outside the project location, such as extreme conditions of weather, while internal variability can be caused by various sources of unstable motivation, workflows and quality issues causing rework [22]. Construction variability presence influences the possible increase in the production rate by improving bottleneck processes [19].

Previous literatures in engineering management studies have revealed that the two major noteworthy instruments for controlling changeability work process are unpredictability caused by rework which can be limited using effective observing and quality review [23, 24]. Walsh and Sawhney [25] saw that workflow process levelling approaches deliver steady outstanding burdens for exchange of temporary workers and workflow process levelling in construction productivity can be connected in different courses.

Bashford and Sawhney [26] carried out an investigation and observed that in the activity-based model, an equally spaced schedule is upheld for all activities. In order to tackle these problems of unnecessary rework and ineffective time, certain management techniques have been evolving which is known as "time and motion study" but are now termed work study [27].

Shirowzhana [28] investigated the use of a Wi-Fi-based positioning and communicating system for indoor positioning with radio communication systems called the Voice Communication and Locating System (VCLS). The VCLS was integrated with BIM and GIS for enhancing labour productivity in the construction industry. The GIS and the BIM were used for displaying the positions of mobile devices for tracking the position of a worker in a proposed site. The study revealed that VCLS has the ability of tracking the estimated locations of workers in 3D environment and GIS and VCLS are valuable software that could improve the communication efficiency and quality of interaction on the infrastructure site.

2.1 Techniques used for measuring time and motion study

There are three major methods used in the measurement of time and motion study (work study) in construction productivity [29]:

1. Visual recording method

2. Physical recording method

3. Interview and questionnaire method

2.1.1 Visual recording method

Visual account is a strategy utilised in observing the execution of specialists on destinations persistently, and it may be considered as a pioneer in the use of visual chronicle method for efficiency checking determinations [29]. Parker and Oglesby [30] saw that this procedure includes the utilisation of photography time-slip with the application of a camera concentrated on a work territory and a photo is taken at precise time interims extending from 2 to 8 minutes. This strategy has a few advantages over others giving right, enduring and unquestionable records that can be valuable to teach determination of construction claims and contract contradictions, assessing workforce skills performance and evaluating the productivity of construction methods. This technique is related with a few burdens, for example, creating distress between the labourers being continually watched, lacking inadequate coverage area and the prerequisite for various cameras on huge destinations.

Another kind of visual recording approach is the utilisation of tape recording framework in estimating workforce abilities execution. It has related elements of a task yet catches all or most concurrent exercises considering less eyewitnesses nearby. One of the burdens of this strategy is discovered that the procedure is a costly technique; it requires taught watchers to be available on location to monitor and record the work finished by specialists. The rudimentary methodologies utilised in this execution are activity sampling, visiting the site every day and consistent observation. Sampling activity strategy is otherwise called sampling work [30].

Activity sampling technique can be portrayed as a sporadic recognition methodology that incorporates seeing a little degree of assignment events that is adequately tremendous to have a numerical consequence. Regardless, it includes an eyewitness who may stroll around the project site and record workforces' events, it is used to pass on a gadget for describing how the pros commit their time on attempted site, and it is occasionally expected that work accomplished, or yield is identified with the measure of period spent on direct work [30]. Activity sampling technique has been used widely and tested by several researchers, and important results were presented in many technical articles; one of the main importance of the method is that several workers can be concurrently observed by an observer on site, while the main drawback accompanying this method is the assumption that the amount of direct time spent is related to the outputs [31].

An investigation carried out on the US plant construction using an extensive analysis of the data collected by activity sampling established there was no relationship among the direct working time spent and the outputs. The investigation found that work sampling technique does not differ between actual and busy work and decided that direct labour cannot be used to estimate productivity [31].

Another type of activity sampling technique is the daily visit technique which is also known as intermittent observation method that encompasses the observer collecting data from site visit on a day-to-day basis [32]. According to Naoum and Hackman [32], finished site work or construction is visually checked by the observer and, if appropriate, marked on drawings; this technique has numerous importance, and if building sites are in near closeness to one another, the viewer can screen numerous sites throughout the same day.

Significance of this strategy is that it evades the turmoil joined by workforces being watched and checked and underpins in maintaining a decent relationship among the watcher and site labourers. The information assembled by this procedure relies upon the exactness and precision of the information given by the site specialists; the steady perception strategy can be isolated into perception by direct technique and work study technique.

2.1.2 Physical recording technique

Direct observation technique can also be described as a physical method of observation that consists of a researcher physically observing the site for the total duration of the operational day and the observer concentrating his devotion on a gang of skilled craft members and recording direct and contributory time spent on the work and the period that is not consumed at work such as late starts and early quits, attending to personal needs and breaks [33].

This technique offers a correct data and is very advantageous in determining the time input distribution used to accomplish certain outputs. The major disadvantage of this method comprises dissatisfaction and mistrust between the workers or group being constantly monitored, which might cause incorrect or inflated productivity, and also, on big project sites, it might necessitate more than one observer to efficiently observe the tasks rendering this method expensive [34]. There are three parallel approaches to physical recording technique, namely, work study, method study and work measurement.

- Work study

The term "work study" considers the parallel strategies of method study and work measurement, which by an efficient method of examination and enhancement try to get the most ideal utilisation of human and materials assets [27].

The work study strategy is another related procedure like the immediate perception technique; notwithstanding, it shifts in the season of checking by the watcher on location; this technique, perception identifies with the work grouping of the undertaking analysed (British standard glossary, BS 3138). Work study can be all around characterised as an association benefit dependent on those techniques; particularly work measurement and method study strategy contemplate what is utilised when exploring human work in all background, and this can prompt the precise examination of the considerable number of capitals and variables which aggravate the viability and economy of the condition being inspected, so as to impact improvement [32].

Work study bolsters the parallel method study and work measurement strategies, which is a methodical technique for acquiring the most ideal utilisation of material and HR. It is a standout among the most possibly helpful apparatuses of the management. Conventional men can accomplish extraordinary outcomes by limiting squandered exertion and time and setting an appropriate standard of execution.

- Method study

Method study measures the adequacy of the strategy and work technique procured to convey frameworks of examination designed for the headway of efficiency measures and surroundings of working [35]. The aim of method study is to evaluate the best method of doing work in order to recommend the best efficient technique of production. Although the detailed analysis may become more complex, the basic procedure is summarised in **Figure 2**.

The initial approach to this method is to first select the type of work that will impact the overall productivity of the skills workforce. The factors to be considered during the selection are economic, technical or human. Economic factors can include "bottlenecks" that can influence other activities. For example, form working on reinforced concrete-framed buildings, the operations can involve lots of labour/plant or the transportation of materials through a long distance like excavation of long hauls and repetitive work like building. There are always technical issues, but these issues must be resolved by a consultant specialist [27].

- Work measurement

A work measurement complements method study in order to derive the maximum benefits from a systematic study of work activities. While method study is related to the approach in which work is performed, work measurement is interested with the human resources involved in the job. Although work measurement and method study are interdependent, they cannot be separated from each other.

The mental approach of work study is primarily the application of common sense, and assessment of the expected time needed to carry out a specific task by work measurement, because of the special technique employed, is mainly the field of the trained and experienced expert. Nevertheless, a project manager must have an adequate understanding of the principle involved if he is to employ work study successfully. Since work measurement seeks to reveal the shortcomings of management and to show up the behaviour of workers, it is often met with resistance so that an understanding of human factors involved is important. It is worth noting

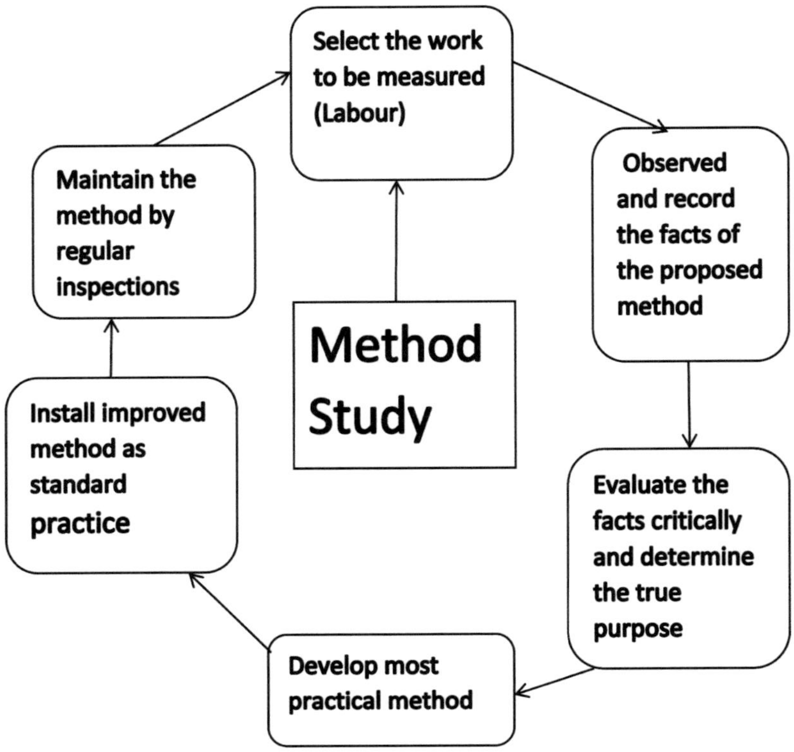

Figure 2.
Method study procedure.

that time study provides a record of a particular operation, and not a check on an individual's performance [27]. The general procedure for work measurement is presented in **Figure 3**.

Work measurement can be carried out by the advanced method. Time study is the basic technique: Timing is usually performed with a watch unless the extremely accurate measurement is required, when devices such as the cine camera or portable tape machine may be used. The watch can be a good wrist or pocket type with a second hand, but a stopwatch is better and more convenient. The purpose of work study is to optimise productivity from the manpower and materials available, and since it is management's responsibility to see that the best use is made of organisation resources, it is, therefore, building managers who must be convinced of the value of work measurement application on building operations [27].

- Work measurement procedure

Work measurement includes structuring a progression of perception structures to incorporate pertinent undertaking data. Site information data is incorporated into the perception shape. Precedent: venture type, perception number, compelling beginning date, space limitation, assuming any, ordinary day by day working hours, and the site management level. Furthermore, a short depiction of the watched task, for example, the technique, divider type, adds up to floor zone, and the quantity of stories will likewise be incorporated into the form. At long last, the qualities of each watched exchange are featured as presented in **Table 1**: work measurement form. The fundamental motivation behind these structures is to reliably record the basic efficiency parameters of data sources and their related yields for the different aptitudes or exchanges and to reflect, to an expansive degree, the genuine conditions on destinations.

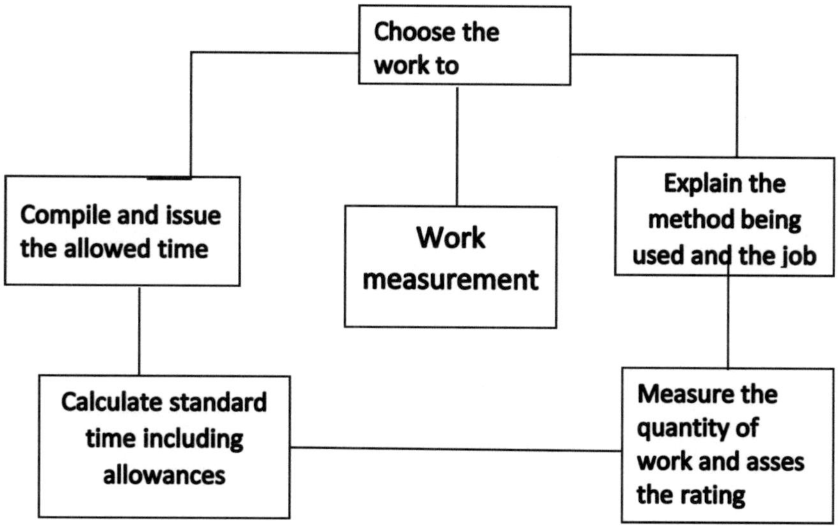

Figure 3.
Work measurement procedure.

Table 2 presents a typical analysis of work measurement data using the simple linear regression technique to investigate the relationship between the cycle numbers of the repetitive productive work input. The dependent variable is the skills worker recorded observed time productivity. A significant level of 5% was used to determine the relationship between skills worker inputs and cycle numbers by substituting the recorded observed time into the linear regression model equation:

$$Y = \alpha + \beta X \tag{3}$$

From the regression equation, α and β indicate the intercept and the slope of the linear regression model. The slope and the intercept are estimated; thus

$$\beta = (n\Sigma xy - \Sigma x\Sigma y)/(n\Sigma x2 - (\Sigma x)\ 2) \tag{4}$$

$$\alpha = \bar{Y} - \beta\dot{X} \tag{5}$$

where y are the man-hours and x are the cycle numbers.

In the regression equation above, α and β were used to calculate the regression model for the skills productivity as presented in **Table 2**.

As depicted in **Table 2**, the slope and the intercept were estimated; thus

$$\beta = (n\Sigma xy - \Sigma x\Sigma y)/(n\Sigma x2 - (\Sigma x)\ 2), \text{and } \alpha = \bar{Y} - \beta\dot{X} \tag{6}$$

In order to measure the significant of the skills (blockwork) productivity, the authors employed the application of the learning curve theory utilising the straight-line unit model as presented in **Table 3**. The mathematical model for the straight-line learning curve is

$$Y = TI \times (\times)\ b \tag{7}$$

where y = cost and man-hours; T1 = cost, man-hours or time necessary to perform the first unit; x = cycle number of the unit; and b = slope of the learning curve, which is calculated as

$$b = InS/ln2 \tag{8}$$

Work measurement data collection form								
Firm:		Study no.:				Date:		
Gang no.:		Start time:						
No. of skilled:	No. of unskilled:	Finish time:						
No. of operators in gang:		Total observed time:						
Contract duration:		Wall thickness:				100 mm		150 mm
Type of building:	Bungalow	225 mm				Other		
Storey building:	Other	Wall height:				0–1.5 m		
No. of floors:		1.5–2.1 m				2.1–3.0 m		
Weather condition:		Above 3.0 m						
		Observer:						
Element description		**R**	**WR**	**OT**	**BT**	**AL**	**ST**	**Remark**
Discharge/loading of material								
Mixing of mortar								
Laying of blocks								
Pointing								
Others (specify)								

R, rating; WR, watch reading/cumulative; OT, observed time; AL, allowance; ST, standard time; BT, basic time.

Table 1.
Typical work measurement form [2].

S/N	LN man-hours	LN cycle no.	C	D	E	F	G	H	I	J	K	M	N	O
	Y	X	XY	X^2	Y^2	$n\sum XY$	$\sum X \sum Y$	$n\sum X^2$	$(\sum x)^2$	H-I	F-G	$\beta = K/J$	$B\dot{x}$	$\alpha = \bar{Y} - \beta\bar{X}$
1	6.10	—	0	—	37.2100	9579.295908	9591.644112	4209.966516	3754.35602	455.6104957	−12.348204	−0.0271	−0.063871109	6.08464
2	6.04	0.6931	4.186324	0.4804	36.4816									
3	6.03	1.1098	6.692094	1.2317	36.3609									
4	6.01	1.3862	8.331062	1.9216	36.1201									
5	6.01	1.6094	9.672494	2.5902	36.1201									
6	6.03	1.7917	10.80395	3.2102	36.3609									
7	6.12	1.9459	11.90891	3.7865	37.4544									
8	6.04	2.0794	12.55958	4.3239	36.4816									
9	6.05	2.1972	13.29306	4.8277	36.6025									
10	6.02	2.3025	13.86105	5.3015	36.2404									
11	6.04	2.3978	14.48271	5.7494	36.4816									
12	6.03	2.4849	14.98395	6.1747	36.3609									
13	6.04	2.5649	15.492	6.5787	36.4816									
14	6.04	2.6390	15.93956	6.9643	36.4816									
15	6.00	2.7080	16.248	7.3333	36.0000									
16	6.01	2.7725	16.66273	7.6868	36.1201									
17	6.04	2.8332	17.11253	8.0270	36.4816									
18	5.98	2.8903	17.28399	8.3538	35.7604									
19	6.01	2.9444	17.69584	8.6695	36.1201									
20	5.98	2.9957	17.91429	8.9742	35.7604									
21	5.98	3.0445	18.20611	9.2690	35.7604									
22	5.99	3.0918	18.51988	9.5592	35.8801									
23	5.98	3.1355	18.75029	9.8314	35.7604									
24	5.98	3.1781	19.00504	10.1003	35.7604									
25	6.01	3.2189	19.34559	10.3613	36.1201									
26	5.98	3.2581	19.48344	10.6152	35.7604									

S/N	LN man-hours	LN cycle no.	C	D	E	F	G	H	I	J	K	M	N	O
	Y	X	XY	X^2	Y^2	$n\sum XY$	$\sum X \sum Y$	$n\sum X^2$	$(\sum x)^2$	H-I	F-G	β = K/J	$B\dot{x}$	$\alpha = \bar{Y}-\beta\dot{X}$
$\sum$	156.54	61.2728	368.4345	161.9218	942.5226									

Table 2.
Regression model for skills productivity.

S/N	LN man-hours	LN cycle no.	C	D	E	F	G	H	I	J	K	L	M	O	P	Q
	Y	X	XY	X2	n∑XY	∑X∑Y	n∑X2	(∑x)2	E-F	G-H	β = I/J	Ẏ	Ẋ	Bẋ	α = Ẏ-Bẋ	S = 2b*100
1	6.1000	—	—	—	9579.2959	9591.6441	4209.9665	3754.3560	−12.3482	455.6105	**−0.0271**	6.0208	2.3566	−0.0639	**6.0846**	**98.1389**
2	6.0400	0.6931	4.1863	0.4804												
3	6.0300	1.1098	6.6921	1.2317												
4	6.0100	1.3862	8.3311	1.9216												
5	6.0100	1.6094	9.6725	2.5902												
6	6.0300	1.7917	10.8040	3.2102												
7	6.1200	1.9459	11.9089	3.7865												
8	6.0400	2.0794	12.5596	4.3239												
9	6.0500	2.1972	13.2931	4.8277												
10	6.0200	2.3025	13.8611	5.3015												
11	6.0400	2.3978	14.4827	5.7494												

S/N	LN man-hours	LN cycle no.	C	D	E	F	G	H	I	J	K	L	M	O	P	Q
	Y	X	XY	X2	n∑XY	∑X∑Y	n∑X2	(∑x)2	E-F	G-H	β = I/J	$\dot{Y}$	$\dot{X}$	B$\dot{x}$	α = $\dot{Y}$-B$\dot{x}$	S = 2b*100
12	6.0300	2.4849	14.9839	6.1747												
13	6.0400	2.5649	15.4920	6.5787												
14	6.0400	2.6390	15.9396	6.9643												
15	6.0000	2.7080	16.2480	7.3333												
16	6.0100	2.7725	16.6627	7.6868												
17	6.0400	2.8332	17.1125	8.0270												
18	5.9800	2.8903	17.2840	8.3538												
19	6.0100	2.9444	17.6958	8.6695												
20	5.9800	2.9957	17.9143	8.9742												
21	5.9800	3.0445	18.2061	9.2690												
22	5.9900	3.0918	18.5199	9.5592												
23	5.9800	3.1355	18.7503	9.8314												
24	5.9800	3.1781	19.0050	10.1003												
25	6.0100	3.2189	19.3456	10.3613												
26	5.9800	3.2581	19.4834	10.6152												
Σ	156.5400	61.2728	368.4345	161.9218												

Source: Bureau of Labor Statistics [3], Ugulu and Allen [12].

Table 3.
Learning rate of skills productivity.

where S is the learning rate, which is defined as the percentage reduction in the unit input, i.e. cost, man-hours or time, as a result of doubling the number of units completed. Eq. 8 can be rewritten as

$$S = (2^\wedge b) * 100 \quad (9)$$

2.2 Questionnaires and interviews

Questionnaire and interview techniques involve collection of information through interviews from workers and supervisors in the construction industry [36]. The two types of surveys that are mainly used by observers are:

1. The craftsmen's questionnaire survey

2. The foreman delay survey

- The craftsmen's questionnaire

The craftsmen's questionnaire survey is an intermittent perception strategy that is utilised for assessing the management productivity and to recognise difficulties impacting skilled workers' output and motivation. This procedure includes a normal premise; the labourers are asked to give a guess of the loss of time on site, rank the seriousness of the difficulties and prescribe system to these difficulties, and therefore, managers would have the capacity to distinguish difficulties activated by deferrals and their plausible administrative suggestions to these difficulties. This strategy produces work joy and inspiration between the specialists as it communicates the feeling of association in the enhancement of the job instead of being only a device for its achievement [37].

The inconvenience of this system is because information gathered together are normally founded on the labourer's memories and estimates as opposed to correct and point-by-point ongoing data.

Shortcoming originates from the measurement that workforces are not in a site to quantitatively perceive the reasons of general delays as they are not regarding the organisation, which lessens the individual technique and inclined to incorrectness, and work may likewise be irritated when workforces are named upon to fill the forms in classification for namelessness purposes; also, the methodology ends up being troublesome and dreary when every representative needs to fill a form.

The investigation conducted by Chang and Borcherding [37] recommended a new method which is related to the activity sampling method. In this method workers are selected randomly and are requested to fill the questionnaire form based on their utmost current events; this technique leads to better-quality accurateness and abridged work interruption as workforces are near to the observer at their various areas of work to complete the surveys form on the spot.

- The foreman delay survey

The basic principle of the foreman delay survey method is to assist foremen, being closer to their work in order to determine and calculate the losses due to time at the close of each day at work with acceptable correctness [38]. This technique involves the foremen being requested to fill a list for day-to-day delay account in the project site, and the obtained information is measured, and a summary of time lost is shown stipulating all the stages of site supervision, together with the foremen who contributed to the data.

This method generates job happiness and motivation for foremen and their teams as they increase in value the fact that their decisions are respected by site management. Furthermore, the foreman delay surveys do not consider the availability of data on the efficiency of work approaches in a job and the capability of the labour force or the level of productivity attained; also, it offers a reasonably priced technique of finding input times activities.

3. How can infrastructure skills productivity be improved?

Previous researchers [12, 14, 39] on improving infrastructure productivity observed that infrastructure productivity can be improved through any of the following:

1. Effective supervision and monitoring
2. Material management and supply chain management
3. Project front-end planning (loading) and work face planning
4. Training/learning and certification of workforce
5. Labour management and relations

- Effective supervision and monitoring

In effective supervision and leadership area of improving construction productivity, professionals in the construction industry should adhere to the following:

- o Ratio of labour to supervision of 1–8 to 1–20. Should not exceed 20
- o Oversight with experience and authority
- o More formal process to enforce company failures
- o Accountability of scope, time and cost
- o Availability of all materials for construction

- Material management and supply chain management

Material management and supply chain managers should ensure the following:

- o Ensure that all materials are available in a timely manner.
- o "Look ahead" material management and logistic plans.
- o Provide more resources, for example, material availability, land surveyors, construction supervisors and contractors.
- o Provide the right materials at the right time and ensure accountability.
- o Build relationships with suppliers.
- o Implement material management and controls in advance.

- Project front-end planning (loading) and work face planning
    - Front-end planning is the method of developing enough strategic information which the project team can address the scope of the project and requirements that allow successful execution of the project, defining long-term plan, objectives and scope rather than jumping into design and quick construction.
    - If long-term goals and planning are more concretely defined, rework can be limited. Define scope of project and reduce conflicting scopes.
    - Plan projects in detail before starting.
- Training/learning and certification of workforce
    - Training programs for both foreign and local labour help companies improve productivity in the following ways:
    - Encourage more organisational training for lower-level supervisors who are directly responsible for people on the tools.
    - Provide training and mentoring for frontline supervision.
    - Provide training for project management field/plant personnel.
    - Ensure proper training on the job to infrastructure worker to enhance experience.
    - Infrastructure trades promotion versus office/clerical alternative education.
    - If resource from other countries is used, then foreman must speak the same language.
    - Investing in apprentice training is necessary.
    - Provision of proper programs for mentorship will improve both work performance and training effectiveness.
- Labour management and relations can be improved through:
    - Better relationship with workers because when management treat workers better, they perform better.
    - Provide attractive incentives, education and skilled labourer retention.
    - Foster a productivity culture productivity with an aim to get above 65%.

4. Conclusion

This chapter featured the systems utilised in measuring productivity in the construction industry. There are three noteworthy techniques that are possible for measuring construction skills productivity. The first identifies visual recording strategy that includes the use of consistent record of the labourer execution output on sites. The second is the physical recording procedure which includes an immediate surveillance system that incorporates a qualified onlooker really watching the site for the full length of the working day; the eyewitness or observer concentration is on a group of skilled workers who record the time spent on direct work. On enormous sites, more than one onlooker might be expected to productively watch the task of work utilising work study, method study or work measurement. The third measuring system is the utilisation of surveys and interviews which include data assembling through a meeting from supervisors and workers working in the construction industry. Finally, the chapter concluded that infrastructure productivity can be improved through effective supervision, material management and supply chain management, project front-end planning (loading) and work face planning, training and certification of workforce and labour management and relations. The above different strategies featured and talked about with illustrations are utilised in the measurement of infrastructure skills productivity.

Conflict of interest

The authors declare that there is no conflict of interest.

Author details

Rex Asibuodu Ugulu[1*], Stephen Allen[2] and Andrew Arewa[3]

1 Department of Quantity Surveying, Federal University of Technology Owerri, Nigeria

2 School of Construction Economics and Management, Built Environment, University of Salford, UK

3 Quantity Surveying and Commercial Management, Coventry University, Coventry, UK

*Address all correspondence to: rexugulu@gmail.com

References

[1] Organisation for Economic Co-operation and Development. 2009

[2] Jarkas AM. Critical investigation into the applicability of the learning curve theory to rebar fixing labor productivity. Journal of Construction Engineering and Management. 2010;**36**(12):1279-1288. DOI: 10.1061/(ASCE)CO.1943-7862.0000236

[3] Bureau of Labor Statistics. Labor Productivity and Costs. 2018

[4] European Association of National Productivity Centres. Productivity: The High Road to Wealth. Brussels: European Association of National Productivity Centres; 2005

[5] American Association of Cost Engineers. International Recommended Practice 10S 90, Cost Engineering Terminology. Morgantown: AACE International; 2013

[6] Allmon E, Hass CT, Borcherding JD, Goodrum PM. U.S. Construction Labor Productivity Trends, 1970-1998. Journal of Construction Engineering Management. 2000;**126**(2):97-104. DOI: 10.1061/(ASCE)0733-9364

[7] Bureau of Labour Statistics. Labor Productivity and Costs. 2006. Available form: www.Bls.Gov/LPC/Faqs.Htm#P01

[8] Akindele OA. Craftsmen and labour productivity in the Swaziland construction industry. In: CIDB 1st Postgraduate Conference University of Witwatersrand Port Elizabeth. South Africa: Construction Industry Development Board (CIDB); 2003

[9] Aki P, Harri H, Maila H. Productivity and performance management—Managerial practices in the construction industry. International Journal of Performance Measurement. 2011;**1**:39-58

[10] Tangen S. Demystifying productivity and performance. International Journal of Productivity and Performance Management. 2005;**54**(1):34-46

[11] Bohme T et al. Causes and mitigation for declining productivity in the Australian mid-rise residential construction sector. Built Environment Project and Asset Management. 2018;**8**(3):253-266. DOI: 10.1108/BEPAM-10-2017-0097

[12] Ugulu RA, Allen S. Using the learning curve theory in the investigation of on-site craft gangs' blockwork construction productivity. Built Environment Project and Asset Management. 2018;**8**(3):267-280

[13] Tanko BL et al. An implementation framework of value management in the Nigerian construction industry. Built Environment Project and Asset Management. 2018;**8**(3):305-319. DOI: 10.1108/BEPAM-09-2017-0078

[14] Forsythe P, Brisland R, Sepasgozar S. Measuring Installation Productivity on Panellised and Long Span Timber Construction. Vol. 2016. Metrics: Permalink. Open Access. Australia: Forest & Wood Products Australia; 2016. pp. 1-93

[15] Attar AA, Gupta AK, Desai DB. A study of various factors affecting labour productivity and methods to improve it. IOSR Journal of Mechanical and Civil Engineering (IOSR-JMCE). 2012;**1**(3):11-14

[16] National Institute of Standards and Technology, National Standards. National Institute of Standards and Technology. 2011

[17] Abdel-Wahab M, Vogl B. Trends of productivity growth in the construction industry across Europe, US and

Japan. Construction Management and Economics. 2011;**29**(6):635 44

[18] National Research Council. Industry Productivity Analysis. Washington, DC, USA; 2009

[19] Arashpour M, Wakefield R, Blismas N. Improving construction productivity: Implications of even flow production principles. CIB world building congress 2013. In: International Council for Research and Innovation in Building and Construction. Netherlands: Rotterdam; 2013

[20] Doloi H, Iyer K, Sawhney A. Structural equation models for assessing impacts of contractor's performance on project success. International Journal of Project Management. 2011;**29**(6):687-695

[21] Hwang BG, Zhao X, Goh KJ. Investigating the client related rework in building projects: The case of Singapore. International Journal of Project Management. 2014;**32**(4):698-708

[22] El-Adaway IH. Insurance pricing for windstorm-susceptible developments: Bootstrapping approach. Journal of Management in Engineering. 2012;**28**(2):96-103. DOI: 10.1061/(ASCE)ME.1943-5479.0000088

[23] Kraft E, Molenaar K. Fundamental project quality assurance organisations in highway design and construction. Journal of Management in Engineering. 2014:04014015. DOI: 10.1061/(ASCE)ME.1943-5479.0000197

[24] Sullivan KT. Quality management programs in the construction industry: Best value compared with other methodologies. Journal of Management in Engineering. 2011;**27**(4):210-219. DOI: 10.1061/(ASCE)ME.1943-5479.0000054

[25] Walsh KD, Sawhney A, Bashford HH. Production equations for unsteady-state construction processes. Journal of Construction Engineering Management. 2007;**133**:254-261. DOI: 10.1061/(ASCE)0733-9364(2007)133:3(254)

[26] Bashford HH et al. Implications of even flow production methodology for U.S. housing industry. Journal of Construction Engineering Management. 2003;**129**:330-337. DOI: 10.1061/(ASCE)0733-9364(2003)129:3(330)

[27] Calvert RE. Introduction to Building Management. 6th ed. London: Construction Industry-Great Britain Management; 1992

[28] Shirowzhana S et al. An integrated GIS and Wi-fi based locating system for improving construction labor communications. In: 34th International Symposium on Automation and Robotics in Construction (ISARC 2017). New South Wales: University of New South Wales; 2017

[29] Fondhal JW. Photographic analysis of construction operations. Proceedings of the American Society of Civil Engineers, Journal of the Construction Division. 1960;**86**(C02):9-25

[30] Parker HW, Oglesby CH. Methods Improvement for Construction Managers. Construction Engineering and Management. New York City: McGraw-Hili Series; 1972

[31] Thomas H. Labor productivity and work sampling: The bottom line. ASCE Journal of Construction Engineering and Management. 1991;**117**(3):423-444

[32] Naoum S, Hackman J. Do site managers and the head office perceive productivity factors differently. Journal of Engineering, Construction and Architectural Management. 1996;**3**(1&2):147-160

[33] Nilson A, Winter G. Design of Concrete Structures. 10th ed. United State: McGraw-Hili Book Company; 1986

[34] Drewin FJ. Construction Productivity. New York, N.Y: Elsevier Science Publishing Co., Inc.; 1985

[35] International Labour Office. Responding to Globalisation: A Decent Work Agenda for the Caribbean in the Context of Regional Integration. Port-of-Spain, Trinidad and Tobago: International Labour Office; 2006

[36] Olomolaiye P, Jayawardane AK, Harris FC. Construction Productivity Management. England: Addison Wesley Longman; 1998

[37] Chang LM, Borcherding JD. Evaluation of craftsman questionnaire. ASCE Journal of Construction Engineering and Management. 1985;**111**(4):426-437

[38] Tucker RL et al. Implementation of Forman delay surveys. Journal of the Construction Division. 1982;**108**(C04):577-591

[39] Jergeas G. Improving Construction Productivity on Alberta Oil and Gas Capital Projects. 2009

Chapter 5

Hospital Energy Resilience

Katerina Vichova and Martin Hromada

Abstract

Many emergencies and crises threaten the whole world. These affect not only the lives, health, and property of citizens but also the subjects and elements of the critical infrastructure itself. Breaking the critical infrastructure system would have a severe impact on the state's security, securing the basic living needs of the population, the health of the people, or the economy of the country. One of the significant sectors of the critical infrastructure is undoubtedly healthcare. It is vital for the health service to be able to perform its function, even in times of crisis. The crisis of naturogenic character ordinarily has the so-called cascade effect, which causes other extraordinary events and crises. An example of this may be windstorm, resulting in a power outage. A significant impact of the power supply outage in healthcare is observed in hospitals. There is currently no assessment tool to set the hospital's readiness for a power outage. The aim of the chapter is to analyze the current state of the crisis preparedness of the hospital.

Keywords: power outage, hospital, energy supply, critical infrastructure, emergency management

1. Introduction

Many extraordinary events and crises threaten the whole world. There were 281 disasters and 61.7 million people were affected by the natural disaster in 2018 in the whole world (based on the International Disaster Database). Next, there were about 1600 storms in 2018. These disasters often cause the power outage. These affect not only the lives, health, and property of citizens but also the subjects and elements of the critical infrastructure itself. Naturogenic or anthropogenic threats could cause it. The aim of each state is to protect its citizens and the necessary infrastructure of the country where systems called critical infrastructure can be considered. Breaking the critical infrastructure system would have a severe impact on the state's security, securing the basic living needs of the population, the health of the people, or the economy of the country. However, the critical infrastructure system must be viewed comprehensively [1].

There are many sectors of the critical infrastructure (see **Figure 1**).

In some cases, the disruption of the critical infrastructure has an impact on the next system. Rinaldi called this as the cascade effects [3]. Next, Rehak examined the cascade effects in the critical infrastructure system [4].

As with any network, a critical infrastructure system has elements with different levels of importance (criticality). The damage, disruption, or failure of an important

Figure 1.
Critical infrastructure sectors [2].

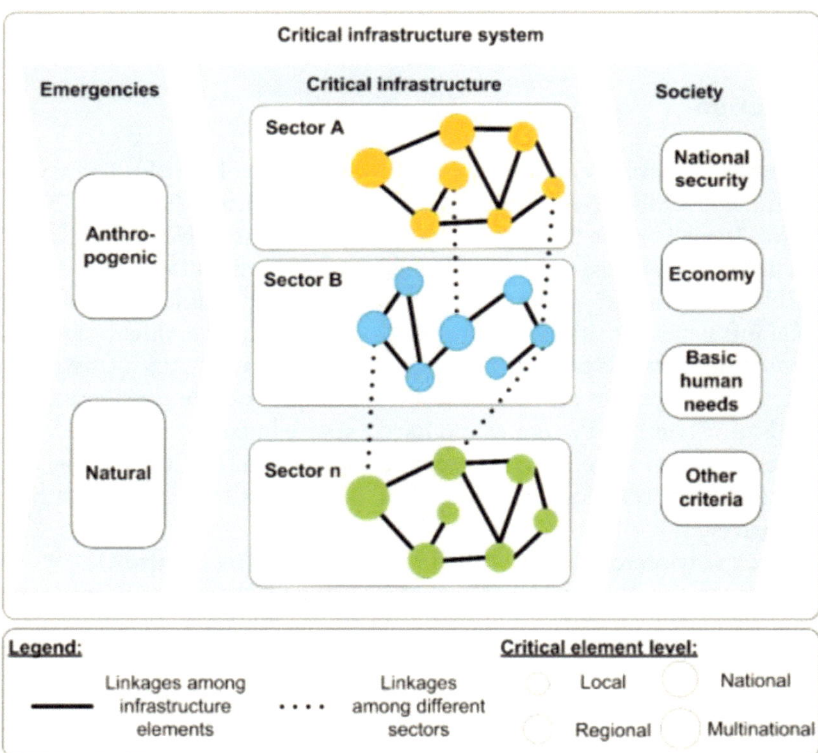

Figure 2.
Critical infrastructure system [4].

(critical) element has a more or less serious impact based on the number and character of linkages that define its level of effect, dependence, or interdependence [4].

The crisis of naturogenic character ordinarily has which causes other emergency — it could be called cascade effect (see **Figure 2**). An example of this may be windstorm, resulting in a power outage. One of the most critical risks for modern science is a large scale power system blackout [5]. Over the decade, electricity plays a major role in the modern day life. Even a momentary power outage can create chaos, revenue loss, and loss of life [6].

It is clear from the list of sectoral criteria that the health sector is also essential for maintaining the state's security and maintaining the functioning of the country. Here, it is necessary to mention the interconnection of critical infrastructure elements. Just breaking the essential element of infrastructure—power outage could result in a cascade effect, namely healthcare.

The aim of the chapter is to analyze the current state of the crisis preparedness of the hospital.

2. Power outage

As has already been noted, naturogenic emergencies and crises may also result in further indications. One of them is a power outage (in some countries we can also find the term blackout). In the Czech Republic, this threat is frequently described as the failure of large-scale electricity supplies, but sometimes we can also find the term "blackout" [7, 8]. In English literature, we can find a replacement for the term "power outage". The term blackout is defined as a total power outage in a large area [9]. Furthermore, this term refers to the moment when the balance between the production and consumption of electricity has been violated, and the security of supply is undermined [7].

Power outage affects not only the lives of people but also the operation of the whole state, especially the economic development of the affected area. The particular feature of the blackout is that the secondary consequences of the outage are many times greater than the effects on the plants for the production, transmission, and distribution of electricity. The cause of this is dominoes that arise from the interconnection of the entire critical infrastructure [8].

It is a predictable factor that causes disturbances on the transmission or distribution system. In this case, the account is also taken of historical events where a power outage was caused mainly by the wind storm. This event could create a cascade effect when one cause gradually raises a series of related events.

Also, the failure of the power supply may be caused by a technical fault. Disturbances, such as transformer fire, can occur both at the power generation points and directly in the transmission and distribution system. If a combination of several serious faults occurs, a significant power outage may occur.

Among other causes, we could include the human factor. In case of concurrence of several negative influences, dispatchers may erroneously assess the situation that may result in a significant power outage. Such conditions are prevented mainly by professionally qualified service personnel and by creating a comprehensive set of safety rules. From the analysis of available data, it is clear that even a terrorist attack may also be a cause of failure. The attack could be made directly, for example, by the destruction of the transformer station, or it could be done through information networks, called the cyberattack.

Finally, there may also be a significant over-flow of energy from foreign grids. An example may be the transport of electricity from power plants in northern Germany to the sampling centers in the southern parts of Germany through the transmission system of the Czech Republic. In the event of a sudden increase in electricity production, along downtime could occur.

Act No. 458/2000 Coll., on business conditions and the execution of state administration in the energy sector and on the amendment of some laws (Energy Act), indicates that an emergency may occur in the electricity grid. A state of emergency means a condition resulting from natural disasters; measures taken by the state authorities for an emergency, a state of state threat, or a state of war; accidents or cumulation of failures in electricity generation, transmission, and distribution facilities; smog situations according to special regulations; terrorist attack; unbalanced electricity grid balances or parts thereof; transmission of a fault from a foreign electricity system; when physical or personal safety is at risk and causes significant and sudden electricity shortages or threats to the integrity of the power system, its security, and reliability in the whole territory of the state, the significant territory or part thereof [10].

Although the company has relied on electricity infrastructure since the early twentieth century, the increasing complexity of infrastructure and its interdependence with other infrastructures such as information and communication technologies makes it more vulnerable. Even if the electricity supply outage is rare, it will occur more frequently and for a more extended period (the result of naturogenic crisis caused by climate change) [11].

Due to the sophisticated security system, the most likely cause of a major power outage is overlapping several primary reasons at one time. Due to the cause of the power supply outage, the rate of renewal of supplies also develops. If there is marked physical damage to the infrastructure, the recovery time will be proportional to the extent of the damage—in days to weeks [12].

Based on the fact that today's world is dependent on electricity, which is used in all areas of life, the power supply outage may have significant consequences. It could, therefore, be assumed that the power outage will occur in all areas—health, public transport, supply, restriction, or interruption of drinking water, gas, heat, outages of mobile operators, Internet network failure, public safety, etc.

2.1 History of the power outage in the world

The whole world is struggling with power outages that last for a variety of periods, affecting states or parts of them, and hence different populations. Power outages are widespread in Asia; however, the low level of robustness of the energy network is taken into account. Nevertheless, even developed countries do not avoid these events, as a cascading effect of an ever increasing number of emergencies and crises.

The northeastern portion of the US and the southeastern part of Canada hit an extensive power outage in 2003. In total, about 50 million people were affected, and in some places, a state of emergency was declared. The cause of this outage was the interconnected problems mentioned. It is said to have been the neglect of maintenance of vegetation along backbone power lines, the inability to detect network problems and communication with neighboring power systems, inadequate training of dispatchers, and lack of backup systems.

This power outage resulted in several events. It was a failure of the mobile telephone network due to overloading by a massive number of calls and a lack of energy to charge mobile phones. Mobile network reliability proved to be quite low and unreliable in a crisis. In some places, there has been a drop in pressure in the water pipes, and there is a risk of the entry of dangerous substances and organisms into the water. The electrified lines were out of service throughout the downtime in the USA. In chemical operation, 140 kg of poisonous vinyl chloride was leaked into the river, and some had to be hospitalized. The mines were trapped by miners who evacuated after the resumption of electricity supplies.

The date of the power outage	Country	Cause	Inhabitants without electricity (millions)
31/03/2015	Turkey	The shutdown of two power plants and simultaneous maintenance on transmission lines (not confirmed)	70
26/10/2012	Brazil	Fire in the substation	53
30/07/2012–31/07/2012	India	The gap between production and dynamically growing consumption (overloaded lines), which has been momentarily exacerbated by unfavorable climatic conditions	670
08/09/2011	USA and Mexico	Operator error and consequent failure of the high voltage line	3
11/03/2011	Japan	The decommissioning of nuclear power plants, after their damage to the tsunami caused by the earthquake	4.4
28/1/2008	China	The snowstorm destroyed the very high voltage line	30
11/11/2009	Brazil, Paraguay, and Uruguay	Short circuit of three transformers due to heavy rains	60
18/08/2005	Indonesia (Java Islands and Bali)	Multiple EC failure to eliminate 2700 MW of the power	100
27/09/2003–28/09/2003	Italy	Storm	56
14/8/2003	USA and Canada	Short branches of the tree and consequently a weak solution to these initiation faults	50
20/02/1998–27/03/1998	New Zealand	Repeated failures on the high voltage cables	1

Table 1.
History of the power outage in the world [author].

In the same year, Italy and Switzerland suffered a significant power outage. It affected about 56 million people. Again, these were cumulative causes that resulted in a significant power outage. The backbone line between Switzerland and Italy was heavily overloaded, and there was a shortage of surrounding vegetation. The operators had 10–15 min to solve this situation, which they could not solve. After the failure of the second backbone line, there was a cascade effect and the interruption of all other lines [7].

The following table shows the history of the significant power outage in the whole world (see **Table 1**).

Table 1 shows the significant power outage from 1998 to 2015. As can be seen, the largest population without electricity was in 2012 in India. This outage hit 670 million of the inhabitants.

2.2 History of the power outage in the Czech Republic

Even though the Czech Republic is using efficient instruments of crisis management in the electricity sector, its requirement ever is taken into account that this event may occur. After the devastating storm Kyrill in 2007 and Emma in 2008, we must admit that [13]. In this respect, critical infrastructure equipment is a back-up powerhouse with its fuel supply, which is capable of covering the consumption of essential offices, hospitals, etc. [7]. Hospitals are a critical infrastructure sector where it is necessary to operate as soon as possible restored.

The failure to supply electricity of the same magnitude as the one mentioned in the world has not yet been recorded in the Czech Republic. There were only local power outages or power line interruptions. These are several, frequently caused by extraordinary events and crises of a natural character.

On the verge of the power outage was the power industry declared an emergency on 24 July 2006 in the Czech Republic. It was not a typical power outage regarding a fatal impact on the consumer—no households have been interrupted by the supply of electricity in the Czech Republic. However, due to the announced regulatory levels, large consumers have had to limit their collection [14]. The reason for this outage was the previous rupture between the Hradec and Etzenricht substation. This line was replaced, but on July 24, when the substation was re-connected to an unexpected outage at the Diviča substation, this increased the demand from the Czech Republic to Austria. The cascade effect then caused further outages [14].

Another reported power outage, this time with an impact on the population, had a cause in the Kyrill storm, which was moved across Western and Central Europe and also hit the Czech Republic in 2007. The CEZ Energy Group has declared a state of emergency in the Czech Republic as a result of this storm. Due to the consequences of the wind turbine, 27 percent of CEZ Group supply points were found without electricity, which includes more than a million households. This condition was caused mainly by the fall of trees and branches into power line, thunderstorm, and other weather influences [15].

Emma was struck by Europe and the Czech Republic a year later. Again, it caused the fall of trees to power lines, which resulted in up to 920,000 households without electricity. The Herwart storm came as a result of the fall of trees to power lines in the Czech Republic 10 years after Emma's. CEZ Group has declared a calamity situation in eight regions. Without electricity, there were 500,000 subscribers. CEZ Group registered approximately 14,000 households without electricity [16].

There was also an intense storm of Fabianne in 2018. This storm caused power outages, primarily because of the trees grabbed on the line. There were 160,000 households without electricity.

From the history of large-scale power supply outages, it can be observed that the greatest cascade effects caused by naturogenic phenomena have the greatest effect on the outage.

2.3 Power outage in the hospitals

This part of the chapter deals with the history of power supply outages with impact toward the hospitals. A list of selected events according to the time horizon in selected parts of the world is presented.

An earthquake hit the islands of Lombok and Bali in Asia in July and August 2018. During this disaster, 98 people died. The earthquake intensity was on the seventh Richter scale. This catastrophe has caused landslides and as a consequence of the downstream cascade effect, power outage [17]. The main hospital in Tanjung in the north was critically damaged and evacuated. Employees prepared a provisional hospital—about 30 beds were in the shade of the trees and in the tent on the playground to have a place to treat the wounded. Other hospitals were overcrowded, and some patients were treated in car parks [18].

Tropical storm Leslie hit Europe in 2018. This storm hit mainly the southern part of Europe. The most affected were Portugal, France, Spain, and the island state of Mallorca. The storm caused a flash flood in these states. The hospitals were flooded, and thousands of people in France were evacuated. A similar situation was also in Portugal, which was also hit by flash floods [19].

Another was the Irma Hurricane in the US. It was the strongest hurricane in the Atlantic in recorded history. The storm arrived on the Barbuda coast in September 2017. Its wind speed was 185 miles per hour for 37 hours. More than 10 million people have stayed in Florida without electricity. More than 5 million Florida residents were evacuated [20]. Some hospitals were affected by Hurricane Irma. Hospital staff decided to dismiss patients who were able to be in home care. Other patients had to be evacuated to other hospitals. In some hospitals, urgent income has been interrupted.

Hurricane Sandy arrived on the US coast in October 2012. It was a post-tropical cyclone. The Hurricane began as a tropical wave in the Caribbean and quickly, in just 6 hours, turned into a tropical storm. It turned into a hurricane during October 24th. The total deaths reached 285, including 125 deaths in the United States. Hurricane caused damage of $ 62 billion and $ 315 million in the Caribbean. More than 7.5 million people were without electricity [21].

Hurricane Sandy caused floods to hit hospitals, causing power outages in New York. Five hospitals were forced to evacuate patients. The evacuation was also required due to power outages. Other hospitals were evacuated due to the flood of cellars. The aggregates that were placed here ceased to work.

These events are just a selection of some naturogenic disasters that have had an impact on the hospitals. A naturogenic disaster did not cause the following power outage. It was a power outage that lasted for 6 weeks in 1998. The city of Auckland was hit in New Zealand. A technical failure caused this outage. The initial assumptions of the downtime were for 1 week and then the estimates of the energy company's employees were specified per month. The number of banking data centers, stock exchanges, central city offices, customs and immigration, inland revenue, internal affairs, social care, Auckland City Council, central police, significant hospitals, University campus and technical institute, television and radio stations, hotels, and more was affected. Although many of these buildings have generators, various switching failures have caused problems and subsequent outages. Some institutions have tackled a power failure by purchasing aggregates from Australia or Poland [22].

The Czech Republic has not met the failure of electricity supply with the impact to the hospitals of such magnitude.

3. Crisis management and healthcare

As mentioned above, the security threats to the Czech Republic, but also the whole world have the character of naturogenic or anthropogenic. These may result in the power outage.

Act No. 458/2000 Coll. regulates energy in the crisis (Energy Act) and the decrees of the Ministry of Industry and Trade (Decree No. 80/2010 Coll., Decree No. 344/2012 Coll., Decree No. 225/2001 Coll.). Electricity for the population will be delivered according to the appropriate level of regulation, tripping, and frequency plan based on the specific situation [10].

The Ministry of Health of the Czech Republic is responsible for crisis management in health care. To that purpose, a crisis preparedness workplace is set up [23].

The Charter of Fundamental Rights and Freedoms states that everyone has the right to health protection. Citizens have the right to free healthcare and medical aid by public (health) insurance under the conditions laid down by law [24].

The Ministry of Health of the Czech Republic states that its role in crisis management is to ensure framework conditions for the provision of health care at the crisis by creating and enforcing state health policy [25].

Crisis preparedness in the health sector is defined as the ability of health care providers, and health care facilities to provide essential health care to the population locally relevant to administrative unit for emergency and crisis in the continuity of medical policy for the provision of health care by qualified personnel [25].

In the introduction of the chapter, the impact of crises on state security and threats to its infrastructures was mentioned. This subchapter describes crisis management in the hospitals.

Healthcare facilities are spaces for the provision of health services [26]. However, this concept includes not only hospitals but also therapeutic institutions, spa health facilities, general practitioners, and others.

For this research, a healthcare facility—a hospital was selected. A hospital is a medical facility licensed to provide health care, has many beds, an organized medical team of the requisite qualification, and offers continuous nursing services. Hospitals are divided by the number of beds to small (up to 700 beds) and large. They can also be classified by type of ownership into hospitals owned and administered by the state, public hospitals administered by regions, towns and municipalities, private hospitals or nonprofit hospitals, and hospitals based on a business principle [27].

The term healthcare is also defined, which can be considered a multidisciplinary field. Healthcare is modern medicine that represents a complex process that accumulates human knowledge from a wide range of disciplines, not just science, but also technical and social sciences. The collected data and acquired knowledge are thus used both for theoretical and scientific purposes, that is, for the development of the discipline as an academic discipline, as well as for the practical decisions of both political and managerial disciplines [28]. From the above definitions of health care, it is evident that there are also other disciplines, one of which is also computer science. It interferes with healthcare not only in the form of electronic patient cards but primarily as a national health register. Nowadays, information support for crisis management is also being used. The objective of information support is to simplify, streamline, and accelerate crisis management with the help of information systems.

As has been mentioned, healthcare is a multidisciplinary field, which also involves other interests. This work attempts to connect three branches, namely health, informatics, and crisis management.

4. Critical infrastructure and healthcare

As mentioned above, the objective of each state is to protect its citizens and the basic infrastructures of the country where systems called critical infrastructure can be considered. Disruption of a critical infrastructure system would have a significant impact on state security, the security of the basic living needs of the population, the health of people, or the state's economy. The Czech Republic defines industry criteria for identifying industries, sub-sectors, and critical infrastructure elements. Healthcare is also one of the critical sectors of critical infrastructure.

With the increasing insertion of new information and communication technologies into all areas of life, resource consumption, and the maintenance of living standards, new threats arise not only for individuals but also for the economy, the state, and society. Threats can arise from not only organized groups and terrorist and criminal organizations, but also manifestations of hostilities of countries, natural disasters, or competing struggles for necessary raw materials. Specific area of society's life more dependent on, critically reliant on another area, the more severely the infrastructure sector is causing the functionality or limitation of the feature of a particular sector [29]. Government Regulation No. 432/2010 Coll. criteria for identifying critical infrastructure elements were introduced in the Czech Republic. This government regulation divides the criteria into cross-cutting and sectoral. Sectoral criteria are defined by the following sectors:

- energy,
- water,
- food and agriculture,
- health,
- transport,
- communication and information systems,
- financial market and currency,
- emergency services, and
- public administration [30].

It is clear from the list of sectoral criteria that the health sector is also essential for maintaining the state's security and maintaining the state's functionality. It should be noted that this is a hospital with a total number of acute beds of at least 2500. However, no condition is fulfilled by any hospital in the Czech Republic.

Because of the need to cope with the crisis in hospitals, it is essential that hospital emergency care must be implemented. In this case, it should be noted that the critical infrastructure referred to is directed to healthcare. In healthcare, as in other areas, eight critical areas are defined (see **Table 2**). In healthcare, they have their specific features and commodities. All of them are essential for healthcare, and their activities are under crisis management [31].

1.	Electricity, gas, and fuel
2.	Supply—drugs, lingerie, food, and water
3.	IT—medical documentation, internal contact, and contact with superiors and other external organizational units
4.	Management
5.	Transport of persons and materials—medical and supportive
6.	Insurance and finance—sharing personnel, health service, and costs
7.	Special substances—medicines, artificial nutrition, and special and basic medical supplies

Table 2.
Critical area of the hospital [31].

5. Supply of healthcare facilities at the time of power outage

Czech Standard ČSN 33 2140 entered into force on electrical wiring in rooms for medical purposes in 1987. This ČSN was valid until 2015. Subsequently, ČSN 33 2000-7-710 was introduced for low-voltage electrical installations; part 7-710 deals with special-purpose devices in particular premises—medical areas. It should be noted that this current ČSN is valid for newly built medical facilities.

For medical premises, a power source must be installed which, in the event of a faulty mains supply, for a specified period and at a predetermined switch-over time, provides power to devices divided by a power source with different switching times.

Power supplies with a switching time of up to 0.5 s including—in the event of a voltage failure on one or more phase wires in a switchboard, a power supply must be used to provide power to illumination of operating tables and another necessary lighting, such as endoscopes, at least after 3 h, and the resumption of voltage must be within 0.5 s [32].

Power supplies with a switching time of up to 15 s—safety lighting and other devices (see below) must be connected within 15 s to a power source capable of supplying power for at least 24 h when the voltage on one or more mains power supply mains security purposes will be reduced to less than 90% of the nominal value for more than 3 s [32].

Power supplies with a switching time of more than 15 s—power supplies for other electrical equipment of medical equipment that do not meet the requirements listed above and are required for health services may be connected to the power supply automatically or manually. This power supply must be capable of delivering power for at least 24 h. These electrical devices can be sterilizing devices; building equipment such as heating or air conditioning, ventilation, building, and waste disposal facilities; cooling device; kitchen furnishings; and battery chargers [32].

Lightning escape routes are considered as safety lighting; illumination of exit signs; all wiring (including switchboards with main switchboards) of rooms with safety and additional safety sources; rooms where essential services are provided; group 1 healthcare facilities; and group 2 medical facilities [32].

Among other devices, we can include selected fire lifts; ventilation systems for smoke extraction; people search systems; medical electrical appliances used in group 2 medical areas which are intended for surgical or other applications of vital importance; electrical devices for the delivery of medical gases, including compressed air, vacuum supply, and anesthetic gas discharge system, as well as their monitoring devices; and fire detection, fire alarm, and fire extinguishing systems.

This standard provides a classification of safety circuits for healthcare facilities (see **Table 3**).

Class	Interruption
Class 0—without interruption	Power is provided automatically without interruption
Class 0.15—very short interruption	Power supply automatically up to 0.15 s
Class 0.5—short interruption	Power supply automatically up to 0.5 s
Class 5—normal interruption	Power supply automatically up to 5 s
Class 15—middle interruption	Power supply automatically up to 15 s
Class >15—long interruption	Power is automatically provided for more than 15 s

Table 3.
Classification for the interruption [32].

Medical area	Group			Class	
	0	1	2	≤0.5 s	>0.5 s ≤ 15 s
Massage room	X	X			X
Bed room		X			X
The birth hall		X		X	X
ECG, EEG, and EHG room		X			X
Endoscopy		X		X	X
Investigation room or nursing room		X		X	X
Urology		X		X	X
Radiological room		X			X
Hydrotherapy		X			X
Physiotherapy		X			X
Anesthesia			X	X	X
Operating room			X	X	X
Operational preparation room			X	X	X
Operational gypsum room			X	X	X
Postoperative room			X	X	X
Catheterization room			X	X	X
Intensive care room			X	X	X
Angiography			X	X	X
Hemodialysis		X			X
Magnetic resonance		X	X	X	X
Nuclear medicine		X			X
Room for premature babies			X	X	X
Intermediate care unit			X	X	X

Table 4.
Medical room and their classification [32].

The classification of importance may vary for each site. In this case, the highest safety requirement obligation is taken. It is possible to refer to the following table (Annex B of the ČSN), which contains the classification of healthcare facilities regarding safety circuits (see **Table 4**).

Table 4 shows the classification of healthcare facilities by group. The author considers the operating rooms, the postoperative room, intensive care units, the delivery room, and the premise for premature babies to be the essential premises regarding electricity supply.

Hospitals are among healthcare facilities whose rapid and efficient services play an essential role in reducing disaster mortality rates [33]. A critical part of crisis management is the search for and mitigation of the risks to the population, depending on assistance and care in the healthcare facilities [34].

Therefore, hospitals should be designed and built to be able to deal effectively with all kinds of crises [35]. One of the problems of the World Health Organization (WHO) is the disaster preparedness of the hospitals [36]. There is still no standard and valid tool for assessing disaster preparedness in the hospitals [37]. Disaster managers need accurate and useful tools to assess disaster preparedness for hospitals. However, there is no standardized and comprehensive instrument for this purpose [38]. Based on the authors' quotations and the analysis carried out, it can be stated that there is still no evaluation system to determine the hospital's preparedness for the crisis.

6. Methodology

In this chapter, four scientific methods were used. Firstly, it was the analysis, which we used for the analysis of the hospitals. Secondly, the method comparison was used, which compares the results from the analysis. Thirdly, the induction method was used, where this method serves to examine the fact of creating a hypothesis from the points obtained. Finally, the heuristic analysis of the preparedness was used.

The heuristic analysis of the preparedness was developed for the evaluation of the hospitals. Based on this assessment, we will get an accurate idea of the weaknesses and strengths of the assessed hospitals.

This analysis was divided into four categories. The hospitals were assessed from the point of view—emergency water supply, emergency food supply, emergency energy supply, and other emergency supply.

Emergency power supply—this category evaluates the preparedness of the hospitals for emergency energy supplies—the ownership of energy supply replacement units. However, these aggregates are fuel-dependent, and it is, therefore, necessary to assess fuel supply to the hospital. It deals with the area of contractual fuelling, its gas station, etc.

Based on the above analysis, an assessment was performed using Eq. (1):

$$HP = \frac{(R + H)}{2 \times H} \times 100\% \quad (1)$$

where HP is hospital preparedness, R is sum of results (obtained points), and H is the number of assessed heuristics.

The evaluation methodology consisted in assigning a response to each question answered in the form of valuation from a predefined set of values (−1 = does not agree, 1 = agree; 0 = partly agree, and blank field if the problem is not relevant).

7. Results

This part of the chapter is focused on assessing the crisis preparedness of the hospitals in the area of emergency energy supply.

The assessment method was proposed—heuristic analysis of the preparedness. This analysis was used for assessing 13 hospitals in the Czech Republic. The Czech Republic is divided into 14 regions. Each region has a different number of hospitals. As a rule, each region has one faculty hospital. However, there are exceptions; for example, the Zlín Region does not have a faculty hospital. There is only a regional hospital. There are also district hospitals, private hospitals, and specialized hospitals. The evaluated hospitals were from five regions in the Czech Republic. Based on the mentioned introduction of this paper, the focus is on emergency energy supply. The following figure shows the results of the 13 hospitals (see **Figure 3**).

Figure 3 shows the emergency energy supply of the hospitals in the Czech Republic. There is a differentiation of the emergency energy supply of the hospitals. As can be seen, the best-evaluated hospital from the point of view is South Moravian Hospital 1. On the other hand, the worst evaluated hospital is in Central Bohemian Region 1. The type of hospital could cause this differentiation.

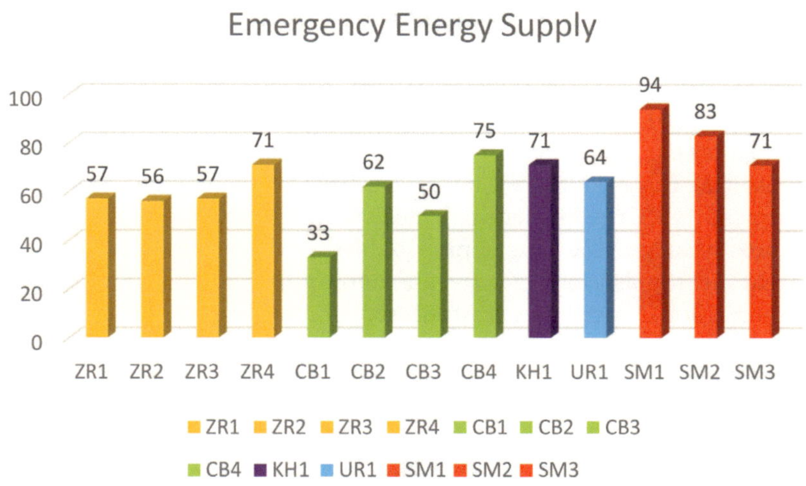

Figure 3.
Emergency energy supply of the hospitals [author].

It was in the readiness of the hospitals to make sure that we saw the supply of electricity and that we saw significant differences. And this fact enabled us to define the primary aim of our research and this chapter.

Each hospital has to deal with the external issue of emergency fuel stocks to address individually. There is currently no valid legislation to ensure this. Some hospitals address this situation through a contractual agreement with fuel suppliers (ČEPRO or other contractors); other hospitals own a gas station where they have sufficient supplies even for a more extended period of the power supply. On the contrary, there are also hospitals that do not solve this problem at all and do not have a contractual agreement with any supplier of fuel. If necessary, they use their ambulances to deliver fuels in barrels from gas stations. Here, however, it is essential to take into account the fact that there may be a power failure in the whole region and there will be no possibility to use all the gas stations.

8. Discussion

In general, healthcare and the provision of health care in hospitals play an essential role for each country. The whole world is threatened by a series of events that may be of naturogenic or anthropogenic character. These events may result in a power outage. The power outage just mentioned has a significant impact on the hospitals. It is therefore crucial that hospitals maintain normal operations even during the crisis.

This chapter deals with hospital energy resilience. The relationship between critical infrastructure-health and power outage is mentioned. This point is discussed regarding nomenclature as well as its history and impact on hospitals. In the past, significant power outage impacts on hospitals could be observed.

Subsequently, a heuristic analysis of the hospital preparedness was carried out. This analysis was used to evaluate the strengths and weaknesses and the subsequent comparison of the evaluated hospitals. A total of 13 hospitals were assessed in the Czech Republic. Based on this analysis, we have concluded this issue. This analysis evaluated the different preparedness of hospitals in the field of emergency energy supply. Each hospital is building a different path to this issue. Therefore, it cannot be said that hospitals will be fully operational in the event of a crisis.

Hospitals belong to the healthcare facilities whose rapid and efficient services can play an essential role in reducing disaster mortality rates [33]. One of the problems of the World Health Organization (WHO) is the disaster preparedness of hospitals [36]. There is still no standard and valid tool for assessing disaster preparedness in hospitals [37]. Disaster managers need accurate and useful tools to assess disaster preparedness for hospitals. However, there is no standardized and comprehensive instrument for this purpose [38]. Crisis management deals with Act No. 240 of 2000 Coll., which does not directly address the disaster preparedness of the hospitals. The exact number of hours that fuel supplies for the aggregate supply must have is not specified. There is only ČSN 33 2000-7-710 that was introduced for low voltage electrical installations; part 7–710 deals with individual purpose devices in particular premises—medical areas. This norm says only until when the electricity must be restored.

Based on the authors' references and the analysis carried out, it can be stated that there is still no evaluation system to determine the hospital's preparedness for the crisis. For these purposes, we propose to introduce a hospital evaluation system.

In the last few decades, science has called the algorithm many practical tools to help solve various computer problems. The algorithm is a set of rules that are typical of specific computer calculations or activities [39].

It is assumed that the use of this system will not only be within the hospital but also by the regional office. We will create a Web portal that will illustrate the hospitals in the region. Each hospital will fill in the data that are required to calculate the length of hospital maintenance (or only selected rooms/circuits) and will be stored in this portal. When the power outage occurs, the expected failure interval will be entered. That will make it clear which hospitals are capable of managing the crisis without having to deliver fuel. Otherwise, the request will be sent to external suppliers or the regional authority (depending on the length of the power supply outage) for the fuel supply. All communication would then take the form of data sentences that would be documented. These data sentences would not only serve to send a request but subsequently to acknowledge the receipt of the application, to submit information about the resources assigned and quantity.

9. Conclusion

The aim of the chapter was to analyze the current state of the crisis preparedness of the hospital. There have been an increasing number of crises not only in the Czech Republic but also in the world in recent years. It has an impact on the critical infrastructure of the country and hence on the lives, health, and property of citizens. Information technologies are increasingly affecting citizens' lives and helping in day-to-day activities, but also in solving the crisis. We can, therefore, see the multidisciplinarity of this field, which also affects the area of population protection and crisis management.

The central part of the chapter was an assessment of the current state. We are increasingly confronted with threats of naturogenic and anthropogenic character. These events have a direct impact, and in some cases, they also cause cascading effects. An example may be a windswept wind that can cause trees to fall on the power lines and a subsequent power outage. The impact is already on the mentioned critical infrastructure of the country, including, among others, the health sector. In the hospitals, it is necessary to supply electricity in the required range and quality. Based on previous research, it has been found that hospitals' preparedness in the field of emergency energy supply is at a different level. Some hospitals in the Czech Republic are not prepared to face a long-term power outage and rely on help from authorities and other crisis management bodies.

Besides, there is no comprehensive and uniform assessment system to assess the readiness of hospitals to face a crisis—power outage. At the end, a possible way to solve this situation was proposed.

The results of the research will be used as a form of information support in solving the crisis—power supply shortage in hospitals and also in other medical facilities and other facilities. In particular, the use will be linked to the needs of the hospital and the regional authority and any external contractors.

Acknowledgements

This research was supported by the Internal Grant Agency of the Tomas Bata University in Zlín, under the project IGA/FAI/2019/001 and Department of Security Engineering.

Conflict of interest

The authors have no conflict of interest.

Author details

Katerina Vichova* and Martin Hromada
Department of Security Engineering, Faculty of Applied Informatics, Tomas Bata University in Zlín, Zlín, Czech Republic

*Address all correspondence to: kvichova@utb.cz

References

[1] Rehak D, Novotny P. Bases for modelling the impacts of the critical infrastructure failure. Chemical Engineering Transactions. 2016;**53**: 1-6. ISBN 978-88-95608-44-0; ISSN 2283-9216

[2] Huntsman. Critical Infrastructure Cyber Security Solutions [Internet]. 2019. Available from: https://www.huntsmansecurity.com/industries/critical-infrastructure/ [Accessed: 17 January 2019]

[3] Rinaldi SM, Peerenboom JP, Kelly TK. Identifying, understanding and analyzing critical infrastructure interdependencies. IEEE Control Systems Magazine. 2001;**21**(6):11-25

[4] Rehak D, Markuci J, Hromada M, Barcova K. Quantitative evalution of the synergic effects of failures in a critical infrastructure system. International Journal of Critical Infrastructure Protection. 2016;**14**:3-17. ISSN: 1874-5482

[5] Braun M, Brombach J, Hachmann C, Lafferte D, Klingmann A, Heckmann W, et al. The future of power system restoration: Using distributed energy resources as a force to get Back online. IEEE Power and Energy Magazine. 2018:30-41. DOI: 10.1109/MPE.2018.2864227

[6] Kumar A, Laxmi AJ. Application of intentional islanding algorithm for distributed energy resources in disaster management. 2016. DOI: 10.1109/POWERCON.2016.7753972

[7] Mares M, Rektorik J, Selesovsky J. Crisis Management: Case Studies. Prague: Ecopress; 2013. ISBN: 978-80-86929-92-7

[8] Brehovska L. Blackout. Kontakt. 2011;**13**(1):107-111. ISSN: 1212-4117

[9] Benes I. Blackout. Prague: Cityplan, spol. s r.o; 2008

[10] Energy Act 458/2000. Czech Republic. In: Collection of Laws; 2019. [Accessed: 05 January 2019]

[11] Heidenstrom N, Kvarnlöf L. Coping with blackouts: A practive theory approach to household preparedness. Journal of Contingencies & Crisis Management. 2018;**26**:272-282. ISSN: 1468-5973

[12] Sedlacek J, Vymazal L. Advice for Citizens—Blackout. Brno, Czech Republic: Krizport; 2018

[13] Benes I. Blackout: The Impact on the Population and the Functioning of the State Administration. Prague: Cityplan, spol. s r.o; 2010

[14] Blackout, problem. 3 pol—journal about positive energy. 2015

[15] News. Millions of people without electricity, CEZ and E. ON have a state of emergency [Internet]. 2003. Available from: https://www.novinky.cz/ekonomika/106997-milion-lidi-bez-proudu-cez-i-e-on-maji-stav-nouze.html [Accessed: 10 October 2018]

[16] Economic newspaper. Storm Herwart in the Czech Republic [Internet] 2017. Available from: https://www.domaci.ihned.cz/c1-65932560-bez-elektriny-jsou-v-cesku-desitky-tisic-domacnosti-silny-vitr-zastavil-i-provoz-lanovky-na-jested [Accessed: 07 January 2019]

[17] Nianias H, Graham Ch. Indonesia earthquake: At least 98 dead as tourists evacuated after 'massive' damage [Internet]. 2018. https://www.telegraph.co.uk/news/2018/08/05/earthquake-kills-19-tremors-hit-lombok-bali/ [Accessed: 07 January 2019]

[18] Embury-Denic T. Lombok earthquake: Tourists 'forced to pay to board rescue ships' [Internet]. 2018. Available from: https://www.

independent.co.uk/news/world/asia/lombok-earthquake-latest-indonesia-tremor-quake-dead-victims-tourists-a8480356.html [Accessed: 07 January 2019]

[19] Tahir T, Allen P, Locket J. France flash floods [Internet]. 2018. Available from: https://www.thesun.co.uk/news/7494537/france-flash-floods-devasation-hospitals-roads-evacuations/ [Accessed: 07 Janury 2019]

[20] Irfan U. Superstrom Sandy May Have Long-Term Public Health Impact [Internet]. 2012. Available from: https://www.scientificamerican.com/article/superstorm-sandy-may-have-long-term-public-health-impacts/ [Accessed: 07 January 2019]

[21] Facts about Hurricane Sandy [Internet]. 2018. Available from: https://www.dosomething.org/us/facts/11-facts-for-hurricane-sandy [Accessed: 07 January 2019]

[22] Auckland's Power Outage [Internet]. 1998. Available from: https://www.cs.auckland.ac.nz/~pgut001/misc/mercury.txt [Accessed: 07 January 2019]

[23] Sin R. Disaster Medicine. Prague: Galen; 2017. ISBN: 978-80-7492-295-4

[24] Resolution No. 2/1993. Resolution of the Bureau of the Czech National Council on the proclamation of the list of fundamental rights and freedoms as part of the constitutional order of the Czech Republic. In: Collection of Laws. 2019 [Accessed: 05 January 2019]

[25] Fiser V. Crisis Management in the Area of the Healthcare. Prague: Fire Rescue Service; 2016

[26] Act No. 372/2011 Coll. on health services and the conditions for their provision. In: Collection of Laws. 2019 [Accessed: 05 January 2019]

[27] Vokurka M, Hugo J. Big Medical Dictionary. 4th ed. Prague: Maxdorf; 2004. ISBN: 80-7345-037-2

[28] Janeckova H, Hnilicova H. Introduction to Public Health. Prague: Portal; 2009. ISBN: 978-80-7367-592-9

[29] Richter R. Critical infrastructure. In: Janeckova H, Hnilicova H, editors. Introduction to the Public Health. Prague: Portal; 2009. ISBN: 978-80-7367-592-9

[30] Government Regulation No. 432/2010 Coll. on Criteria Critical Infrastructure Criteria. In: Collection of Laws. 2019. [Accessed: 14 December 2018]

[31] Drabkova J. Healthcare in energy and personnel crisis conditions. In: Janeckova H, Hnilicova H, editors. Introduction to the Public Health. Prague: Portal; 2009. ISBN: 978-80-7367-592-9

[32] Czech Standard. Low voltage electrical installation. In: Part 7-710: Single-Purpose Devices and Special Objects-Medical Devices. Prague: Czech Standards Institute; 2013. ČSN 33 2000-7-710

[33] Arab M, Zeraati H, Akbari HF, Ravangard R. A study on the executive managers' knowledge and performance, and their hospitals preparedness against earthquake events and their relationships at public hospitals. Journal of Health Administration. 2009;**11**(34):7-14. ISSN: 2008-1200

[34] Brehovska L, Nesporova V, Rehak D. Approach to assessing the preparedness of hospitals to power outages. Transactions of VSB—technical university of Ostrava. 2017;**7**(1):30-40. ISSN: 1805-3238

[35] Malekshahi F, Mardani M. Abilities and limitations of crisis management in Shohadaye Ashayer and social

security hospitals of Khorramabad in 1385. Journal of Critical Care Nursing. 2008;**1**(1):29-34. ISSN: 1365-2702

[36] Aradam A et al. Hospitals safety from disasters in I.R. Iran: The results from assessment of 224 hospitals. PLoS Currents Disaster. 2014;**28**(6):1-18

[37] Jenkins JL et al. Review of hospital preparedness instruments for national incident management system compliance. Disaster Medical Public Health Preparedness. 2009;**3**(2):83-89

[38] Heidaranlu E, Ebadi A, Khankeh H, Ardalan A. Hospital disaster preparedness tools: A systematic review. PLoS Current. 2015;**14**(7):1-16

[39] Wroblewski P. Algorithms. Brno: Computer Press; 2015. ISBN: 978-80-251-4126-7

Chapter 6

Bridges: Structures and Materials, Ancient and Modern

Arturo Gonzalez, Michael Schorr, Benjamin Valdez and Alejandro Mungaray

Abstract

Every nation needs the infrastructure to perform all kind of activities related to the improvement and service of the society. Transportation system became part of the infrastructure due its connection between two destinations, using maritime, land, or aerial methods, creating a link for social and economic activity. Bridges are widely used to cross rivers, valleys, and roads, providing a passage with other parts of the land since ancient times to modernity. Each structure has different requirements to cover, such span clearage, traffic flow, geometry and characteristics of the place to build; therefore, a great variety of bridges can be developed. Common materials used on construction are structural steel, reinforced concrete, pre-stressed concrete, or post-tensioned concrete; depending on the structural behavior of each type of bridge, there will be a maximum clear span to cover, which depends directly on the project's budget. There are a variety of loads and environmental conditions that the new and existing structure needs to support effectively, including dead load, traffic, rain, wind, flood, and seismic events, using effective structural design process and techniques; on the other hand, there are long-term deterioration process, such as corrosion, wear, and fatigue, which should be considered on the maintenance process, avoiding additional costs, several damages, and catastrophic failures. Prevention and control of degradation process is achieved by effective maintenance methods applying protection technology such as paints, coating and cathodic protection. The purpose of this chapter is to show a brief review of ancient and modern bridges, including the process of design, material selection, construction, and maintenance.

Keywords: infrastructure, bridges, maintenance, deterioration, construction

1. Introduction

If a society wants all its function working efficiently, it requires many elements that meet its needs. Basic functions such as food, water, electricity and transportation require all types of structures to fulfill its purpose. Infrastructure embraces all the buildings that support the goods and services of the community, with an integral and optimal functioning.

All the services of the society require some type of support to be carried out. An engineer, architect or lawyer needs a space to develop their businesses and support its clients; a merchant needs a highway to transport their products to their final destination; supplying water to any destination requires pipes, pipelines and tanks. There are many examples where each service requires some type of infrastructure.

As a general example, consider the construction of an international airport. The main purpose is the aerial communication of the city, which can be used for traveling, tourism and transportation of goods. Listing all the needs of any airplane, there will be at least the following requirements to fulfill:

a. Landing strip to perform effectively the air traveling.

b. Control tower for monitoring the air traffic flow.

c. Airport building to perform all administrative activities.

d. Hangar for airplane maintenance and storage purposes.

e. Water supply, storage, pipelines and sewage for the entire place.

f. Fuel supply and storage for the airplanes.

g. Land highways and bridges to connect the city with the airport.

h. Electricity towers and electrical station to supply energy.

According the previous example, there are many construction elements and buildings that enhance the aerial communication. The combinations of these structures contribute with the airport to operate efficiently. This type of infrastructure can contribute enabling tourism, transportation, productivity and employment opportunities, increasing the economic activity. Therefore, the infrastructure has a wide variety of structures which can be part of a specific sector and fulfill a simple function.

If any infrastructure element of the airport fails, there will be issues with its functionality; for example, if any bridge is closed due any malfunction or maintenance, the transport of the passengers, goods and services will be affected.

Bridges have a special place in transportation infrastructure due its direct relationship with other places. These structures have the purpose to carry on the traffic loads of the highway, crossing any obstacle and perform an effective communication between two destinations. Since there are many variables to consider in the performance of the bridges, such geometry, span clearance, traffic flow and available materials, there are many options of bridges to choose.

Planning, design and construction process for any bridge looks logical and necessary steps, looking for the good behavior of the structure during any traffic load or resisting flood or seismic events. However, maintenance process guarantee the life of the structure, which applied correctly, will avoid any closure of the bridge and traffic issues.

On this chapter, the main purpose is to focus on bridges as part of land transportation infrastructure, its behavior and the performance during the design, construction and maintenance process.

2. Infrastructure

Basically, all the buildings are part of the infrastructure and fulfill a specific function. According to the American Society of Civil Engineers (A.S.C.E.), there are

16 categories in the infrastructure: Aviation, bridges, dams, drinking, water, energy hazardous waste, inland waterways, levees, parks, ports, rail, roads, schools, solid waste, transit and wastewater [1]. Due to a wide variety structures, the infrastructure can be classified according to its use.

a. Transportation infrastructure

Related to all structures used by the people, products, goods or services to move forward its final destination. Depending on the type of transport, these can be divided into three categories:

i. *Land:* City roads, highways, train rails and bridges.

ii. *Maritime:* Structures used by ships or vessels, for example ports and channels.

iii. *Aerial:* Airports and heliports.

b. Energy infrastructure

Related to all structures intended for the generation and distribution of any types of energy. Depending on the energy type, it can be divided into:

i. *Electricity:* Electrical stations and supply networks.

ii. *Oil and gas:* Refineries and pipelines.

iii. *Alternative energy sources:* Wind towers, nuclear plants and geothermal plants.

c. Hydraulic infrastructure

Structures intended for water distribution and supply, divided into:

i. *Water distribution network:* Related to water supply, distribution and irrigation. Examples include open channels, pipelines and aqueducts.

ii. *Drainage networks:* Related to distribution and storage purposes, including gray water from industry and rain. Examples include sewer and drainage.

iii. *Waste water treatment plants:* Related to structures with waste water cleaning purposes, including removing sediments to biological cleaning process.

iv. *Water storage:* Related to structures to retain water. Examples include dams and storage tanks.

d. Telecommunication infrastructure

Related to all structures intended for telecommunication industry:

i. *Cellphone network:* All structures supporting signal development for cellphone operations, including antennas and signal structures.

ii. *Television, radio and internet network:* Related to antennas for signal distribution via cable wires and wireless signal.

e. Building infrastructure

Related to all structures intended for industry requirements, business community, energy operations and living places.

i. *Industry and business:* Structures used for industry and business operation purposes. Examples include a single or multiple story structures and may be used as offices, machinery, equipment and industrial process.

ii. *Living places:* Intended for population housing, including buildings of single or multiple story levels.

iii. *Basic services:* Structures related to operation services as water distribution, electricity and general supplies.

iv. *Primary services and recreational purposes:* Places designed for city assistance, including fire and police stations, hospitals, schools, theaters and stadiums.

Figure 1 shows a general view of Mexico City and the Vidalta cable-stayed bridge as transportation infrastructure, connecting via highways the whole series of multi-story buildings in the background.

According the previous infrastructure classification, all described structures do not work as isolated buildings. Instead, all elements must be connected looking for the harmony in a society, working effectively and should not affect any building performance.

As an example, there is a hospital that provides all types of medical treatment in an urban area. If the city's electrical stations stop working, the hospital will cease to be fully functional due to lack of electrical power, affecting medical equipment and lighting requirements. Now, if the same hospital has no enough roads for access, its capacity will be very limited, including medical staff and patients.

Figure 1.
View of Vidalta cable-stayed bridge, Mexico City [2].

3. Bridges, a general overview

We recognize that all structures are part of the infrastructure and each one works together. The bridges take a special role, due its function to connect two different points, crossing valleys, rivers, lakes and cliffs.

Bridges are needed on land transportation infrastructure because they connect different points that usually can be inaccessible. If we analyze a single bridge crossing a river, it can have many views, depending on each person's perspective [3]:

a. A person who lives in the city can visualize the bridge as a simple access to schools, parks and theaters, or a simple way to visit a family member.

b. An engineer or architect visualizes the bridge as a way to connect the road with two points of the city, such as hospitals or fire stations.

c. From the business community, the bridge can be viewed as access to different areas for trade, distribution of goods and services.

Depending on the needs of the society to have a bridge, it would be its importance. A bridge that serves as a quick link to recreational parks with a low traffic flow will have less impact than a bridge crossing a large river and connecting two points of the city with high traffic flow.

Taking into account the sentence above, we can realize that bridges are not built arbitrarily; a whole planning should be performed including design, construction, operation and maintenance of the structures. Therefore, for the transportation system, the bridge is a key element [3] according to the following reasons:

a. Capacity control

i. Bridges must comply with traffic flow needs during its life period.

If a bridge has a small number of lanes, narrow sizes or poor spaces, the structure cannot maintain a continuous vehicular flow.

ii. Bridges must comply with required loads during its life period.

This means that the analysis and structural design must take into account all the loads that the bridge must support. For example, if the structure is located on an interstate highway and was not designed to support heavy truck loads, it will have limited vehicle traffic and those trucks will not be able to use the bridge; therefore, these trucks will have to plan an alternate route.

b. High cost for the entire road system

i. Bridges represent a high percentage of road's budget.

If we analyze the construction process of a road and measure the cost per unit distance, the bridges are very expensive compared to the highway.

ii. High cost variability for different bridge geometries.

Depending on the number of lanes required, types of vehicles to be supported, distances and/or clear span to cover, materials and available labor, the cost of the bridges are variable. Proper planning is required to meet the needs and comply with the budget.

c. The bridge as part of the system

i. If the bridge has a failure, the road system fails.

If we analyze the entire road and at specific place, one of the bridges does not work, the vehicular flow will be affected, increasing traffic flow, delays, time lost and the need of alternative route.

ii. If the bridge is not operating, alternate routes will be affected.

When a bridge does not work, people who used the affected road will have the need to use an alternative route, which probably is not designed for a sudden increase of vehicular flow, causing wasted time, greater distances and additional fuel required.

For a successful transportation system, a balance should exist between vehicle volume, supported loads and proper budget. **Figure 2** shows a section of Mexico City's transportation system, showing balance between the number of lanes, signs, bridges and several road accesses.

3.1 Types of bridges: ancient and modern

We can think that all existing bridges with the variety of materials, geometries, loads and designs have always existed. However, the evolution of bridges has occurred within the changing needs of the society since the 19th century, with the improvement of materials, optimization techniques, architectural and structural designs [3].

3.1.1 Ancient bridges

In the beginning, bridges were built with a simple geometry and had very limited uses, because they only covered very short span, such as small rivers. These bridges used basic materials such as wood, ropes and stone.

3.1.1.1 Stone arch bridges

The first bridges that were built based on mathematics methods were the stone arch. The exact construction date is not known, but there are structures built by civilizations such as the Greeks or the Romans, where they used this type of bridges as aqueducts, roads for people walking and carriages.

The stone arch bridges, as shown in **Figure 3**, take advantage of the compressive capacity of the rock due its geometry, supporting its own weight and live loads. These structures are usually robust and each of the arches supports the upper deck. Some examples of this bridge's type are the Segovia Aqueduct, located in Spain and the Pont Du Gard Aqueduct, located in France, both built between the 1st and 2nd centuries.

Figure 2.
View of Mexico City road system [4].

DOI: http://dx.doi.org/10.5772/intechopen.90718

3.1.1.2 Wooden and steel truss bridges

By the beginning of the 19th century, the structures used wood as common material and truss bridges began to emerge (**Figure 4**). These wooden trusses took advantage of the axial stress capacity of the bar elements, creating bridges with longer spans, low weight and enough stiffness to withstand higher loads. Due the large number of geometries that can be created with trusses, there are possible arrangements which the bar elements can have taking advantage of the tension and compression stress capacities. Some examples can be mentioned:

a. Trusses with straight bars

    i. Geometries as Pratt, Warren and Baltimore types.

    ii. Geometries using cables for tension elements.

b. Combination between trusses and arch

    i. Geometries as Wernwag and Burr types.

3.1.2 Modern bridges

In the mid-19th century, with the development industry sector, vehicles and trains entering into circulation using the current transportation system. Therefore, a greater number of roads, railroad and bridges were built to serve the increase of transportation demands.

Figure 3.
Stone arch bridge crossing a small river.

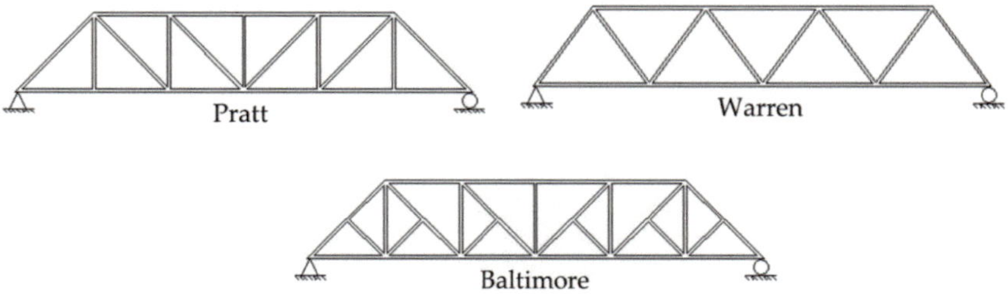

Figure 4.
Geometry types of trusses.

3.1.2.1 Steel truss bridges

Since the increase of traffic flow and weight of vehicles started, the wooden bridges were already insufficient to support these vehicle loads and the structures began to use steel materials. With the structural steel available on construction market, these bridges had a significant improvement, including the increase of spans length covered and supporting higher loads. Steel trusses bridges replaced wooden trusses and began to build bigger structures. **Figure 5** shows an example of this bridge's type.

3.1.2.2 Suspension bridges

The introduction of suspended bridges was an important innovation, due the very large spans length that they can cover. These structures have very large geometries with visual impact on the users, using them as a symbol for the city. Examples of such structures are the Brooklyn Bridge, located in New York and the Golden Gate Bridge, located in California. The overview of the Golden Gate Bridge is shown in **Figure 6**.

The geometry of the suspended bridges consists of two central support towers, the main cables supported between the towers, the secondary cables supported on the main cables, the main deck and supporting girders. As a structure, all cables work as tension elements and support the main deck where the traffic flows.

Due the tension cables, suspended bridges can take advantage of the ability to obtain very large spans without intermediate supports. They are widely used to cross very large rivers where conventional bridges are unable. However, one of the

Figure 5.
Forth bridge, Edinburgh, Scotland [5].

Figure 6.
Golden Gate suspended bridge, San Francisco, U.S.A. [6].

disadvantages of this type of structures is the aerodynamic stability, product of the slenderness relationship between main slab, span clearage and the action of thrust forces produced by the wind loads.

3.1.2.3 Reinforced, pre-stressed and post-tensioned concrete bridges

With the introduction of Portland cement on the market and the development of concrete construction techniques, design theories for reinforced concrete were developed for structures on the early 20th century.

For long span bridges, reinforced concrete bridges based on arch below main deck are used. It basically consists of an arch in the lower section and piers to support the main deck. All the elements of the bridge are working under compression, with the exception of the main deck that works at flexure. These types of bridge take advantage of the material capacity on compression and avoid tension elements.

Figure 7.
Bridge of pre-stressed concrete Girders el Zacatal, México [7].

Most of the existing bridges cover small and medium spans, which are very useful for roads and highways connecting cities. Girder-based structures with simple or continuous supports are widely used for these cases.

Girder-based bridges can be inefficient due the bending behavior of the girder if we compare it with trusses. However, the girders are relatively easy to build and the relationship between cost and benefit makes these types of structures economically competitive. **Figure 7** shows El Zacatal Bridge, located on Mexico and based on pre-stressed concrete girders.

3.2 Structures and functions

Depending on the span to be covered, the traffic flow, the availability of materials and labor, the designer will define the geometry of the bridge. Most of the bridges have short spans and use girders of reinforced concrete, pre-stressed concrete or structural steel. Bridges with intermediate spans use trusses and arches. For very large spans, suspended bridges are the best option.

All types of bridges must have the following qualities:

a. Cover the vehicular flow demand, with enough lanes and/or spaces.

b. Support dead, live and accidental loads.

c. The structure is economically viable.

3.2.1 Structural analysis and design

To comply with the previous qualities, the general statement to ensure safety on the structural design of any bridge should follow the next equation:

$$Resistance \geq Effects\ of\ the\ loads \tag{1}$$

The structural design process includes two general ways to comply with Eq. (1) and develop safety structures [1]:

a. Allowable Strength Design (ASD)

   i. This procedure uses the linear behavior of the materials with a defined yield strength which is located below the ultimate strength.

   ii. Safety is obtained specifying the effects of the loads should produce stresses as a fraction of yielding stress.

b. Load and Resistance Factor Design (LRFD)

   i. This procedure reduces the resistance multiplying a resistance factor φ, usually less than 1; and the load is multiplied by a load factor γ, usually greater than 1.

   ii. Since each load has different levels of recurrence, these factors will vary depending on the load type.

The general way to obtain the stresses depends directly on the applied force, the internal force and the geometry of the structural element [8]. The behavior of each load applied can be listed as follows:

3.2.1.1 Axial stress

Applied to elements with tension or compression forces.

$$\sigma = \frac{P}{A} \tag{2}$$

Where:
σ: Axial stress. Units: lb./in^2 (N/mm^2)
P: Internal axial force. Units: lb. (N)
A: Cross-sectional area. Units: in^2 (mm^2)

3.2.1.2 Direct shear stress

A to elements with direct shear forces.

$$\tau = \frac{V}{A} \tag{3}$$

Where:
τ: Direct shear stress. Units: lb./in^2 (N/mm^2)
V: Internal shear force. Units: lb. (N)
A: Cross-sectional area. Units: in^2 (mm^2)

3.2.1.3 Torsion stress

Applied to elements with torsional moments.

$$\tau = \frac{Tr}{J} \tag{4}$$

Where:
τ: Torsional stress. Units: lb./in^2 (N/mm^2)
T: Torsional moment. Units: lb-in (N-mm)
A: Cross-sectional area. Units: in^2 (mm^2)

3.2.1.4 Bending stress

Applied to elements with bending moments.

$$\sigma_b = \frac{Mc}{I} \tag{5}$$

Where:
σ_b: Bending stress. Units: lb./in^2 (N/mm^2)
c: Distance between neutral axis and external fiber. Units: in (mm)
I: Moment of inertia of the cross-sectional area. Units: in^4 (mm^4)

3.2.1.5 Shear stress due bending

Applied to elements with bending moments.

$$\tau_b = \frac{VQ}{Ib} \tag{6}$$

Where:
τ_b: Shear stress due bending. Units: lb./in^2 (N/mm^2)
Q: Moment of area. Units: in^3 (mm^3)
I: Moment of inertia of the cross-sectional area. Units: in^4 (mm^4)
b: Width of the cross-sectional area. Units: in (mm)

For structures with combination of forces applied at the same times, all stresses are present and interact in the same time. Therefore, the stress combination can be represented with maximum and minimum principal stresses, as shown below:

$$\sigma_{1,2} = \left(\frac{\sigma_x + \sigma_y}{2}\right) \pm \sqrt{\left(\frac{\sigma_x - \sigma_y}{2}\right)^2 - \tau_{xy}{}^2} \tag{7}$$

Where:
σ_1, σ_2: Maximum and minimum principal stress.
σ_x, σ_y: Normal stress on x or y direction due axial or bending forces.
τ_{xy} : Shear stress due direct shear forces, torsion or bending forces.

Units for all stresses: lb./in^2 (N/mm^2)

According to Eq. (7), we should note that on maximum and minimum principal stress, the shear stress is always zero. Now, the maximum shear can be found following the next equation:

$$\tau_{max} = \sqrt{\left(\frac{\sigma_x - \sigma_y}{2}\right)^2 - \tau_{xy}{}^2} \tag{8}$$

In the case of the maximum shear stress, the normal stress is not zero and can be found as follows:

$$\sigma_{prom} = \left(\frac{\sigma_x + \sigma_y}{2}\right) \quad (9)$$

3.2.2 Bridge categories according the location of the main deck

The bridges can be classified according to its size, geometry, main function and structure type. As a general way, the bridges can be divided into three categories:

Category 1: The structure is located below the main deck.

a. Straight trusses.

b. Trusses with arch geometry.

c. Arches with stone or masonry material.

d. Rigid frames.

The geometry of this type of structures allows the user to have a clean view of the road. In addition, most of the structural elements of the geometry have to work under compression stress.

Category 2: The structure is located above the main deck.

a. Trusses.

b. Suspension bridges.

c. Cable-stayed bridges.

For these types of structures, the geometry is fully visible and for large bridges, such as suspended or cable-stayed, the user can appreciate the architecture. Most of the structural elements have to work under axial loads, mainly tension.

Category 3: The structure coincides with the main deck.

a. Girder-based bridges.

    i. Lightened and solid slabs.

    ii. Girders with varieties of cross sections.

These types of structures work mainly under bending and shear stresses. Bridges of this type are the most used for short span.

3.3 Special bridges: Mexicali, Tacoma, Coatzacoalcos, Calatrava Jerusalem

Due to the great imagination of design and construction process, there are a large number of bridges in operation with a wide variety of geometries. Therefore, below are a few examples to show.

3.3.1 Mexicali bridges: solution of the road distributor

The Mexicali road distributor consists of a series of bridges connecting the main roads with the purpose of traffic flow continuity. Basically consists of two reinforced concrete bridges as underpass section and two structural steel bridges as overpass section. This structure is shown in **Figure 8**.

The underpass bridge section, both for vehicular and a railway line, consists of a reinforced concrete slab supported at the ends by retaining walls and circular columns of reinforced concrete supporting the center of the span. The main structure of this bridge section works as continuous girder. The spans are relatively short and the structural slab depth is enough to fulfill the flexure stresses requirements.

The structure also has two structural steel decks as an overpass bridges, consisting of continuous girder and supported by steel columns. The length of the spans is small since they do not exceed 165 feet (50 m). A particular feature of this particular structure is the energy dissipation device located below the girder supports. Mexicali has high seismic activity and the structure needs to withstand the seismic loads with good lateral displacement performance.

3.3.2 Tacoma narrows bridge: a lesson learned

The suspension bridge located at Tacoma Narrows consists of two main structural steel towers supporting a main cable and the main deck is stiffened by two steel girders. A total length of 5905 feet (1800 m) and a span of 2930 feet (893 m) were covered. It was inaugurated in 1940 and became one of the largest bridges in the world.

The main feature of this bridge was the dramatically collapse of the main deck after a few months of inauguration, due to the oscillating movement with the action of the wind flow. These forces were considered for structural design; however, with a much slower wind velocity, the vibration movement increased with enough speed to make the structure collapse. Looking into **Figure 9**, the oscillating movements of the bridge can be observed.

Under research, the main reason for the collapse of the bridge was the concept of resonance, which means, a range of coincidence between the natural frequency of the structure and the frequency of wind thrust loads. The concept of vibration and resonance is not visible easily and many factors influenced on the event:

Figure 8.
Mexicali bridges, road distributor [9].

a. Very high slenderness ratio of the bridge.

b. Use of girders instead of truss as deck stiffener.

c. Obstruction of air flow due the girder itself, causing whirlwind.

After this event, studies on aerodynamics and aero-elasticity topics in the structures increased significantly, developing procedures to simulate these events on structures, including bridges of very large spans.

3.3.3 Coatzacoalcos bridge: Mexico infrastructure

The cable-stayed bridge located in Veracruz, Mexico consists of two main reinforced concrete towers that support the main deck with cable tensors and the slab stiffened by two reinforced concrete girders. The structure has a total length of 3838 feet (1170 m) and a span length of 944 feet (288 m). **Figure 10** shows an overview of the structure.

Figure 9.
Tacoma narrows suspension bridge, under aerodynamic vibrations [10].

Figure 10.
Coatzacoalcos II cable-stayed bridge overview [11].

Figure 11.
Jerusalem cable-stayed bridge [12].

The bridge consists of 15 supports, 14 sections and the main structure. The towers and secondary columns are made of reinforced concrete; the main girder is shaped like a drawer with reinforced concrete and the cable tensors supporting the main deck are made of structural steel. It was inaugurated in 1984 and considered one of the largest structures in Mexico.

3.3.4 Jerusalem bridge: architecture and modernity

This cable-stayed bridge is located in the city of Jerusalem and has a total span of 1181 feet (360 m). The bridge aims to help the city light trail system and the structure consists of a main tower connecting the structural steel slab using 70 steel cables and reinforced concrete supports.

The main feature of this bridge belongs to the architecture and geometry. It was designed by Santiago Calatrava, a world-renowned architect and engineer, and the user can recognize the structure as an unconventional bridge. As shown in **Figure 11**, we can see the special geometry of the main tower and each of the cable tensors, showing a harp shape. In addition, the main deck has a curved form.

The structure was inaugurated in 2008 marking a symbol in the city of Jerusalem. Due to the great height of the main tower and its harp-shaped geometry, the bridge can be appreciated from any place of the city.

4. Bridges construction and materials

Due the large number of variables on the conceptual design of a structure, there is no special formula for determining the best option of a bridge. Many variables come into play, from the experience of the engineers and architects, to the specific needs of the place, such as topography, soil characteristics and materials availability.

There are several models to describe the general process of design, built, operation and maintenance of a bridge in a general way. One of the most compact flowcharts was proposed by Addis [13], shown in **Figure 12**.

The process for any bridge design consists of input data, regulations, design process and results, explained as follows:

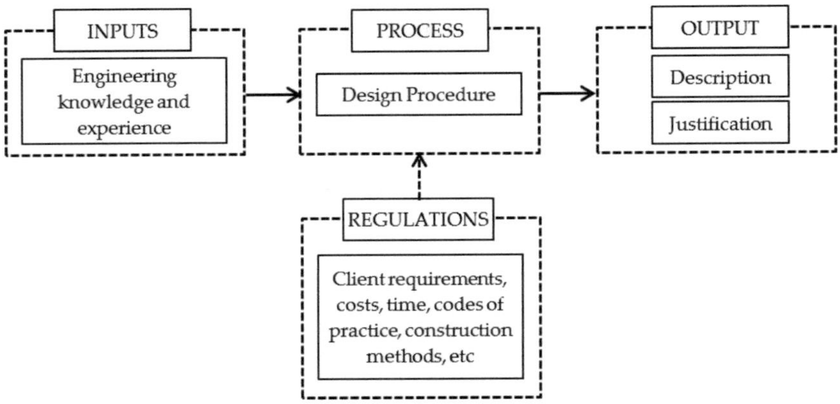

Figure 12.
Model of the bridge design process [13].

4.1 Inputs

All the information required to start the design process of any bridges is placed in this category and can be classified as public and personal. Public information refers to all existing bibliography like books, magazines, publications and software available in the industry. These references should include all topics related to bridges such as material properties, construction process, architectural design and structural design. Personal information refers to the experience acquired by engineers, architects and companies dedicated to the construction industry.

4.2 Regulations

All the rules, restrictions and limitations imposed on the process of creating and design fall into this category. Details such as the budget, the client's guidelines, construction regulation and allowed materials are some of the established rules.

4.3 Output

All the information processed to be able to build any bridge is placed in this category and can be divided into the description and justification of the results. The description refers to all drawings, including architecture, structural, facilities and roads. The justification refers to all technical information that supports the drawings, from structural engineering to budgets.

4.4 Design procedure

In the central part of the flowchart is located the bridge design process, where the input data, regulations and results are interacting together. The design of a bridge implies the imagination of engineers and architects to solve the problem statement, use of the previous knowledge to select the best geometry option and justify the solution with the required calculations.

The flowchart process applies to any type of bridge and can be simple or complicated as required. If we want a successful development of any bridge, there must be a balance between the variables described in **Figure 12**.

Another point regarding the design process of bridge is the selection of the appropriate material and geometry. According to **Table 1**, a recommended bridge's type is shown using geometry, material and span range selection variables [3].

Bridge type	Material	Span range
Slab	Concrete	0–40 ft. (0–12 m)
Girder	Concrete	40–1000 ft. (12–300 m)
Girder	Steel	100–1000 ft. (30–300 m)
Cable-Stayed	Steel	300–3500 ft. (90–1100 m)
Truss	Steel	300–1800 ft. (90–550 m)
Arch	Concrete	300–1380 ft. (90–420 m)
Arch	Steel	800–1800 ft. (240–550 m)
Suspension	Steel	1000–6600 ft. (300–2000 m)

Table 1.
Span lengths for various bridge types [3].

The recommended span range is related directly with budget challenges of each project. As an example, consider the construction of 100 m span length structure which can be developed using a concrete slab and concrete girder, according the recommendations of **Table 1**.

Performing a structural and design of the proposed bridge, we can find the minimum size for the concrete slab and the concrete girders; considering concrete slab, the thickness to support 100 m of span will require a great depth in slab and therefore, a large amount of concrete material will be required; therefore, if we use girders, the amount of material will be less in comparison.

Depending the span range and geometry of the project, the best economical option of bridge selection will be the efficient use of each material mechanical properties, stress-strain relationship and the characteristics of the site.

4.5 Steel bridges

Bridges with steel material can enter into any of each three categories described on Section 3.2.2. Depending on the type of steel to be used, yielding allowable stress of the structural steel can vary between 36 ksi (249 MPa) and 70 ksi (483 MPa). According to the American Institute of Steel Construction [14], common steel alloys are A36, A992 and A572 Grade 50.

Within the steel bridges, the most common geometries are:

a. Straight truss, variable geometry truss or arc-shaped trusses.

b. Cable-stayed bridges.

c. Suspended bridges.

d. Bridges supported by girders.

A steel truss bridge is shown in **Figure 13**, with straight truss at the center of the span and variable height near the column supports. The incremental height on the truss near the columns occurs due an increment axial stress in each truss member. The foundation, anchorage and check slab are made of reinforcement concrete; piers can be made of steel or reinforced concrete, depending the site characteristics.

Steel cable-stayed bridge and suspension bridge with general geometry are shown in **Figures 14** and **15**. Both structures have a main tower supporting the main

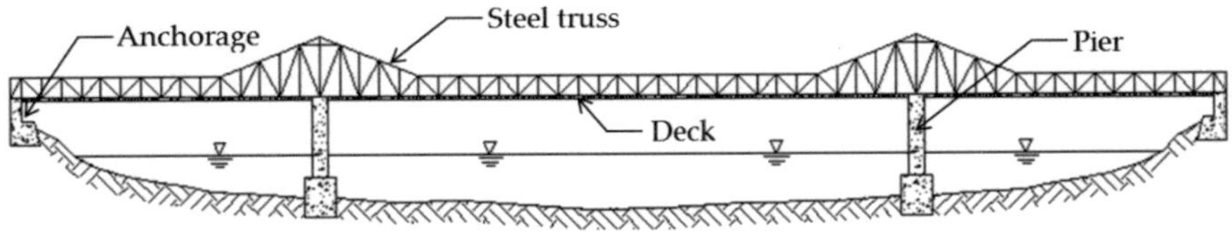

Figure 13.
Steel truss bridges for long span lengths.

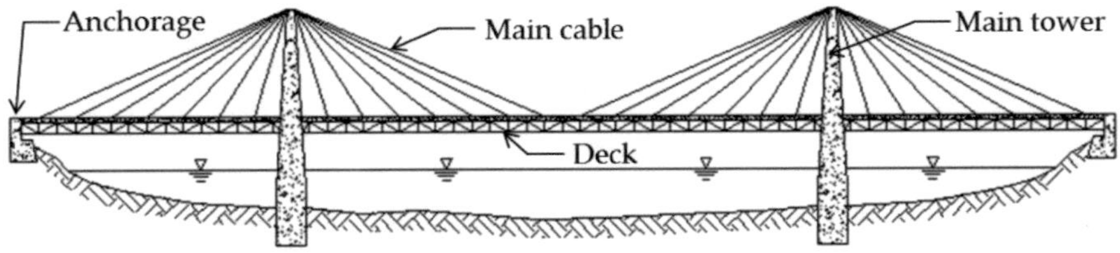

Figure 14.
Steel cable-stayed bridges for long span lengths.

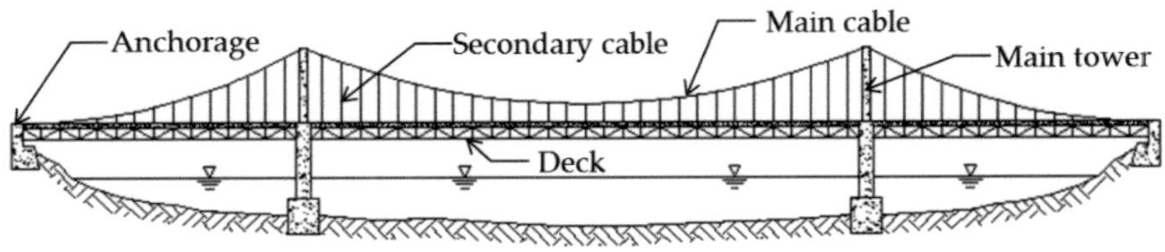

Figure 15.
Suspension bridges for long span lengths.

cables; the difference between these two bridges is the arrangement of the cables. Cable-stayed bridges use a series of cables to support the deck connected directly with the main tower; when the suspension bridges use a main cable supported between the towers and a series of secondary cables supporting the main deck.

For both cable-stayed and suspension bridges, the main deck has a high slender ratio due the long span covered and need additional structural elements to increase the stiffness. Trusses are commonly used to stiff the main deck and allow the wind to flow through these structural elements.

Tension stress is developed by the cables, which are the optimal geometry giving a capacity to increase the span length. Looking into **Table 1**, for span lengths higher than 3500 ft. (1100 m), the suspension bridge is the only economical option to choose.

Bridges supported by steel girders are shown in **Figure 16**. The main deck is the combination of the concrete slab, a wide variety of structural steel beam, piers and anchorage geometries. The steel girders can be simply or continuous beams using hot rolled sections or developed by steel plates.

Steel girders are working with bending stresses, which usually requires more material if it is compared with truss elements. However, according to **Table 1**, these types of bridges can be economical competitive for short and medium span lengths due its easy construction procedures and less time-consuming during installation of the girders. Also, these girders have a great stiffness compared with truss bridges, reducing vibration responses produced by traffic and wind flow.

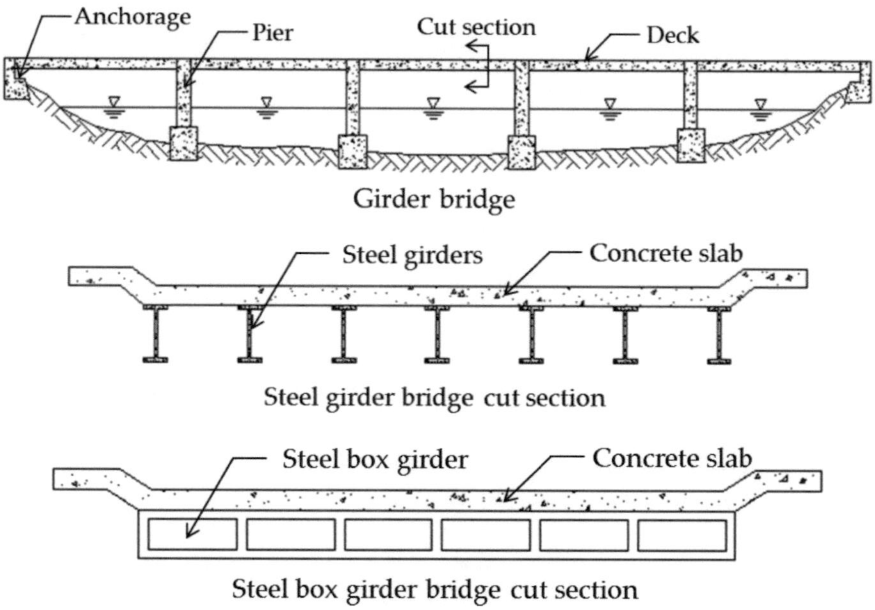

Figure 16.
Steel bridges for short and medium span lengths.

4.6 Concrete bridges

Concrete bridges can be categorized as below or directly on the main structure, as described on Section 3.2.2. According to the American Concrete Institute (A.C.I.), the compression strength of concrete can vary from f'c of 3 ksi (20 MPa) to 7 ksi (48 MPa), depending on cement, water, natural gravel and sand ratios used [15].

There are many advantages of concrete material compared with structural steel, including its capacity to support compression stresses and the availability on construction industry. Tension stresses are carried out by the reinforcement, making a composite structural material.

Within the reinforced, pre-stressed and post-stressed concrete bridges, we can find the following geometries:

a. Arc-shaped concrete below the main deck.

b. Cable-stayed bridges, where the entire structure used concrete except for tensors.

c. Bridges supported by girders.

Arc-shaped concrete bridge is shown in **Figure 17**, which consists of an arc shaped element below all the structure, supporting the piers and the main deck. The concrete arch-shaped element is working mainly by compression stress due its curvature, taking advantage of the material capacity. Piers are working as flexure-compression stress and the main deck is working as shear and bending stress. According to **Table 1**, the recommended span length for structural and economical purposes is 300–1380 ft. (90–420 m).

The principal feature of pre-stressed concrete girders against simply reinforced concrete girders is the increase of the span length without the need of increases the

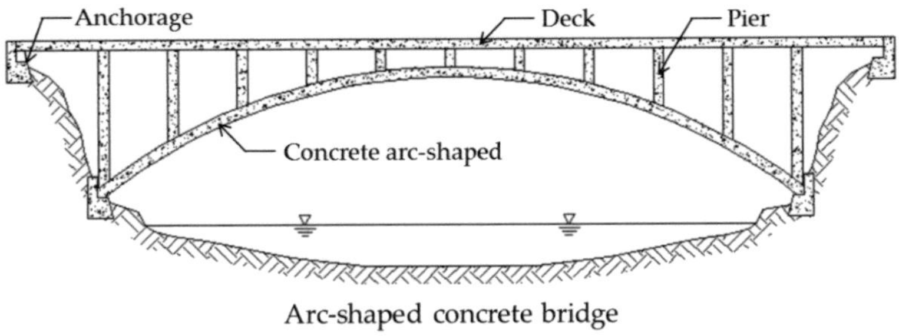

Figure 17.
Concrete bridges for medium span lengths.

beam height, taking advantage of the effective inertia and providing greater stiffness to the bridge. This geometry type is widely used to build bridges across the cities, highways or interstate roads.

According to **Table 2**, there are a wide variety of recommended girders, considering precast pre-stressed or cast-in-place post-stressed concrete with different cross-sectional geometries, taking account the clear span to cover and the material mechanical properties [16].

Each construction procedure have its own benefits; for example, precast pre-stressed girders have the advantage of less time installation consuming and minimum frameworks to use compared with cast-in-place post-stressed girders or cast-in-place slabs, but only can be performed a simple cross-sectional area; by the other hand, cast-in place girders can have any desired cross-sectional geometry, which is adaptable and commonly required on any project.

A concrete girder bridge is shown in **Figure 18**, considering few types of construction procedures and geometries, using the same piers and anchorage.

Cast-in-place reinforced concrete slab or T-beams can be used for small span lengths, as recommended in **Table 1**, and precast pre-stressed I-beams are used for spans lower than 150 ft. (45 m) according **Table 2**. All these types of girders works for bending stress, which limits the span range; however, due its easy construction procedures, are widely used for most common bridges.

4.7 Other materials

Most bridges use structural steel and concrete as main materials. However, there are other materials that can help to complement the structure, depending on some features:

Bridge type	Span range
Precast pre-stressed I-beam	0–150 ft. (0–45 m)
Cast-in-place post-stressed box girder	100–300 ft. (30–90 m)
Precast balanced cantilever, constant depth	100–300 ft. (30–90 m)
Precast balanced cantilever, variable depth	200–600 ft. (60–180 m)
Cast-in-place cantilever segmental	200–1000 ft. (60–300 m)
Cable-stayed with balanced cantilever segmental	800–1500 ft. (240–450 m)

Table 2.
Span lengths for various concrete bridge types [16].

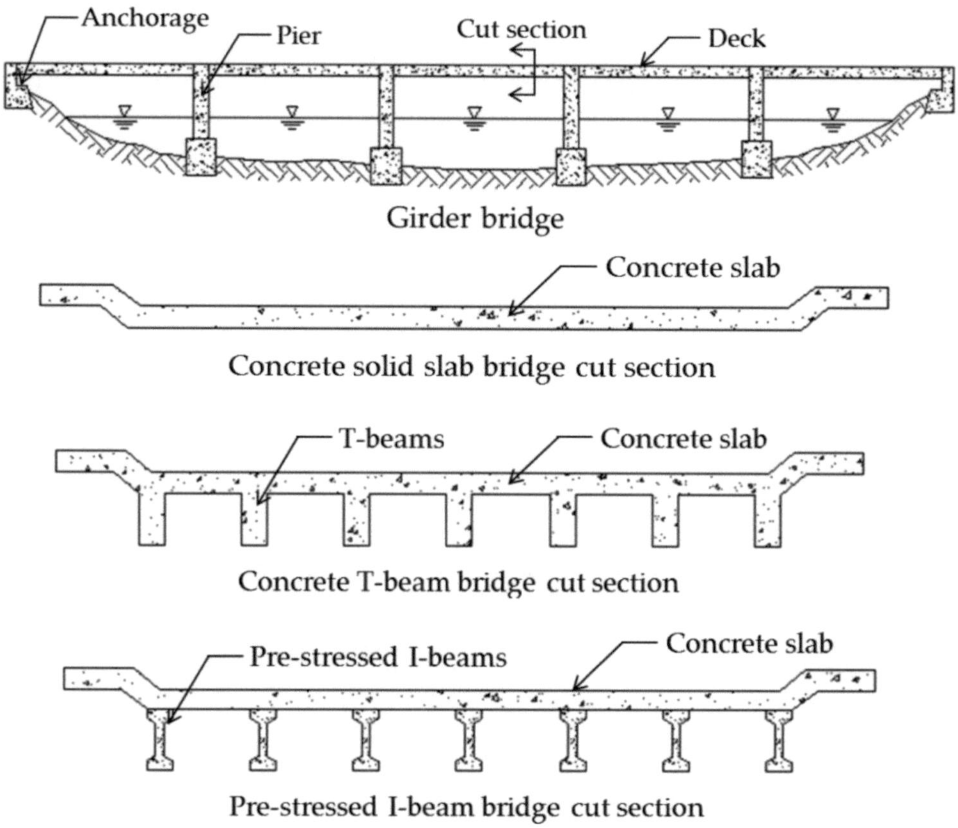

Figure 18.
Concrete bridges for short span lengths.

a. Wooden bridges, used for small crosswalks or where span lengths are short and loads are low.

b. Stainless steel, where it replaces carbon steel parts of the bridge, increasing resistance to humidity and environmental factors.

c. Carbon fibers, used as rehabilitation process and perform capacity improvement of existing structural elements.

5. Maintenance avoids bridge deterioration

The general process for the development of any bridge are described in the flow chart showed in **Figure 12** and includes planning, design, operation and maintenance procedures. To ensure the useful life of the bridge, a maintenance plan must be established, depending on the physical and environmental factors.

According to AASHTO, there are a high variety of loads that the bridge must support and should be considered in the structural design process [17]. These loads are considered as physical factors and can described as follows:

a. Dead loads

Refers to the own weight of the structure, including installations, finishes, bearing surface and all loads that will not have variability over time.

b. Live loads

This type of loads refers to a generalized use of the bridge, this means the traffic flow and people walking, including braking, impact, collision and their dynamic loads. This category includes environmental factors such as rain and snow.

c. Accidental loads

This type refers to an extraordinary event that the structure needs to support, commonly produced by wind and earthquake loads. In some cases, a collision by a ship or a flood event may be considered.

In addition to physical loads, there are other factors that can affect the useful life of the bride, named environmental factors and described as follows:

a. Humidity

This environmental factor affects the chemical composition of iron and steel materials, which in direct contact creates corrosion process and develops a material degradation.

b. Abrasive factors

These factors can affect the material composition due the chemicals reactions by air, water and soil exposures. Each material could have a specific chemical reaction and depends for the levels of exposure.

Dead, live and accidental loads affect the mechanical properties of the structure itself, which results are stresses and deformations. If a load exceeds the capacity of the material, some type of damage will occur, from permanent deformation to crack growth, and the structural element will require a repair.

The environmental factors are associated with the material degradation process, which can results in reduction of the effective inertia. All cases imply the reduction of material resistance and element sizes, developing a stress and strain increase, and therefore, a possible failure.

Proper maintenance avoids possible damages on the structural elements due any physical or environmental factors, and therefore, an increase of the structure expected life.

6. Degradation process: corrosion, wear and fatigue

Since the physical and environmental factors are present in each structure, all materials may be subjected to alterations in their chemical composition, modifying the mechanical and physical properties, shortening the useful life of the structure and requesting any kind of repairmen. These factors with the alteration are listed in this section.

6.1 Degradation due environmental factors: corrosion

Structural steel and concrete reinforcing steel in presence of humidity will have the problem of corrosion, a chemical process involving an electrochemical reaction which occurs due the direct exposure to water, creating rusting and developing the material degradation process.

DOI: http://dx.doi.org/10.5772/intechopen.90718

For reinforcing steel in concrete structures, the corrosion problem can be present when the rebar is exposed, oxidizing the area and develops structural problems.

Figure 19 shows a damaged pier due corrosion of the reinforcement steel, losing the coating and reducing the cross-sectional area, which means a reduction of the mechanical capacity of the element.

On the other hand, if the structural steel is fully exposed to the environment without any humidity protection, the level of oxidation will be present on the entire element and will develop a generalized rusting reaction. **Figure 20** shows a structural steel bridge with all elements damaged by corrosion.

6.2 Degradation due physical factors: wear and fatigue

Wear degradation during the life of a bridge occurs due its continuous use, where the friction is present by physical forces, including the pass of the vehicles over the main deck. These vehicles generate frictional forces when perform braking and accelerating, causing wear on the structure. For bridges where the piers are in contact with water flow, the friction causes degradation.

Usually traffic flow perform low wear degradation over the deck, however, if the road have any defect, will create bumps and wear will be increased rapidly, creating damages. Bridges using simply supported girders will require construction joints between supports; these joints are examples of places were bumps are easily created

Figure 19.
Concrete piers of reinforced steel with corrosion problems [18].

Figure 20.
Bridge of structural steel with corrosion problems [19].

due poor construction procedures. **Figure 21** shows a typical bump problem, which can be avoided using any joint procedure and materials offered in the industry.

Another physical factor that affects the structure is fatigue, caused by the loading and unloading forces due the traffic flow, affecting the stressed elements of the bridge. Fatigue causes degradation on material mechanical properties with each load cycle; this means each vehicle passing over the deck. If each cyclic load produces a stress equal or higher to yielding stress of the material, therefore a large amount of cycles will cause a decrease on the material allowable, allowing brittle failure of the element. **Figure 22** shows a crack developed due fatigue stress loads; notice there are no yielded zones on the beam, only a sudden crack.

6.3 Prevention and protection

Since material degradation process is inevitable for both physical and environmental factors, a few actions must be considered on the design process to control

Figure 21.
Bridge concrete construction joint damaged [20].

Figure 22.
Steel beam with fatigue crack failure [21].

Figure 23.
Steel beam with fatigue crack failure [23].

corrosion, wear and fatigue. Prevention and protection procedures are required, including maintenance process to avoid possible damage.

Corrosion prevention is the best economic way to preserve the structural elements of any bridge and the result is a positive benefit in the useful life of the materials [22]. Acidic corrosive emissions and hydrocarbons, in combination with high humidity accelerate the process of corrosion and degradation. The designer should analyze the type of electrochemical attack that would occur during the life of the structure.

For structural steel, epoxy paints are used to insulate direct contact of water or moisture. For the reinforcement of concrete, should take care of the coating to avoid exposure to moisture. As shown in **Figure 23**, the possibility of install a cathodic protection system should be considered.

Wear degradation process is unavoidable for any surface subjected to friction; therefore, the damage depends of the applied load and affected zone. If a flexible roadway is used, wear degradation is higher and requires additional maintenance compared with reinforced concrete roadway. By the other hand, reinforced concrete piers may have contact with the water flow of the river; frictional forces would be present and wear factor became an issue, requiring additional coating to protect the reinforcing bars.

For fatigue degradation process, the designer must consider the weight of all vehicle types, cyclic loads, loading scenarios and fatigue material properties as a way to prevent brittle failure. All elements should have enough stiffness to avoid high stresses under typical cyclic load cases, therefore the fatigue allowable stress should be greater than the applied loads. Fatigue procedures as Modified Goodman Diagram or Miner's Rule are used [24].

7. Testing and monitoring

To ensure the success of all methods implemented to avoid bridge degradation, a testing and monitoring plan must be established as part of the maintenance procedure. The cost of maintenance plan should be incorporated on the bridge's budget [25].

As a first step, visual review of the bridge structural elements should be performed in a scheduled given time. The girders, piers, connections, cables, deck and materials used must show no damage, such as cracks, corrosion, visible deformations or any variable that indicates a problem. Monitoring techniques are used as a way to measures the loading cycles, cracks or corrosion and prevent any damage on the bridge.

For corrosion mitigation and prevention, the maintenance process must have a plan taking into account the next features:

a. Expected useful life of the bridge.

b. Environmental exposure.

c. Classification of the bridge.

d. Details of corrosion mitigation and prevention methods.

e. Maintenance programs.

If there is no budget on review and monitoring procedures, maintenance will not be a preventive action and becomes corrective, which means higher repair costs and partial or total closure of the bridge.

8. Conclusion and recommendations

The definition of infrastructure includes wide variety of structures, each one with a specific purpose and its function serves the development of the society. All elements of the infrastructure are connected and any issue of an individual part will affect the entire system, slowing down the economic growth.

Transportation infrastructure has the purpose to connect two places using aerial, land and maritime methods, developing a wide variety of structures to comply with its objective. Bridges are part of land infrastructure and are used as link between two places with difficult access using single roads or highways. These types of structures have higher construction costs per mile if they are compared with single roads; therefore, there must be a complete plan to develop the project, which should include the number of lines required to meet traffic flow, vehicle load, site, geometric and budget requirements.

Large bridges exist since hundreds of year ago, where its use was restricted to aqueducts, carriages and road connectors for travelers, using as construction material stone and wood. The introduction of train and vehicles on the industry with the development of structural steel and Portland cement, modern bridges began, increasing load capacity and span length to cover.

There is no specific formula to choose the best option of bridges, due large amount of factors that depends on the structure, as geometry of the bridge, the experience of the construction companies, materials, loads to be carried out, labor available, budget and local site restrictions. A whole process that involves planning, design, construction and maintenance of the structure must be established.

The best economical option for a bridge is the combination of the efficient interaction of geometry and material, taking advantage of tension elements as the main structure. As we can see in **Table 2**, cable-stayed and suspension bridge are the best economical option for span lengths higher than 1800 ft. (550 m), due tension capacity of the steel cables. The span recommendations of **Tables 1** and **2** are a combination of efficient stress capacity due proposed geometry, in alignment of low deformations and material savings, making lower costs and improving budgets.

While planning, design and construction stages of the bridge are carefully studied to ensure its functionality; maintenance is given less importance thinking that the structure will have its useful life without any problems, when reality implies a degradation process, due its use and environmental factors. Maintenance plan is needed to avoid over costs during the life of the bridge.

Wear and fatigue degradation can be carefully studied and analyzed during the design procedure process, which will be present on all structures. However, corrosion degradation depends of the local site environment, mainly humidity and water contact, considered automatically as maintenance plan. Some cases, this maintenance plan is not developed during the design process, and therefore, is not implemented by the time required.

Maintenance plan is always required for a long term useful life of any structure, because degradation process is present all the time. There are issues to be solved if damage due environmental factor is present and maintenance plan was not developed during design process, including budget, repairmen high cost of damaged elements and closure of the bridge. As a conclusion, these issued can be avoided if proper maintenance plan is developed during the design process.

For a successful bridge development, a complete plan should be considered in the entire design process, including the bridge's proposal, design process, construction methods and maintenance program. All variables together will result as a long-term useful life for any structure.

Author details

Arturo Gonzalez, Michael Schorr*, Benjamin Valdez and Alejandro Mungaray
Universidad Autonoma de Baja California, Mexicali, Mexico

*Address all correspondence to: mschorr2000@yahoo.com

References

[1] American Society of Civil Engineers (A.S.C.E.). Infrastructure Report Card; 2017

[2] Picture of Vidalta Cable Stayed Bridge, Mexico City. Available from: https://www.plataformaarquitectura.cl/cl/02-304269/puente-vidalta-mexpresa

[3] Barker RM, Puckett JA. Design of Highway Brides. 3rd ed. Wiley; 2013

[4] Picture of Mexico City Road System. Available from: https://periodicocorreo.com.mx/abren-manana-el-distribuidor-vial-en-la-glorieta-santa-fe/

[5] Picture of Forth Bridge, located on Edinburgh, Scotland. Available from: https://www.railway-technology.com/projects/forth-rail-bridge-firth-scotland/

[6] Picture of Golden Gate Bridge, San Francisco. Available from: https://www.diariodelviajero.com/america/puente-golden-gate-datos-y-curiosidades

[7] Picture of El Zacatal Bridge, Mexico. Available from: http://www.sintemar.com/es/rehabilitacion-de-pilotes-en-puente-el-zacatal

[8] Goodno BJ, Gere JM. Mechanics of Materials. 9th ed. Cengage Learning; 2018

[9] Picture of Mexicali Bridges. Available from: https://www.flickr.com/photos/buelna/4386351666

[10] Picture of Tacoma Narrows Suspension Bridge of 1940. Available from: https://www.txstate.edu/news/news_releases/news_archive/2015/November-2015/TacomaNarrows110315.html

[11] Picture of Coatzacoalcos II Cable-Stayed Bridge. Available from: http://construyendometas.blogspot.com/2009/05/puente-coatzacoalcos-ii.html

[12] Picture of Jerusalem Cable-Stayed Bridge, designed by Calatrava. Available from: https://www.dezeen.com/2015/12/03/a-z-advent-calendar-chords-bridge-jerusalem-santiago-calatrava/

[13] Addis W. Structural Engineering: The Nature of Theory and Design. Ellis Horwood; 1990

[14] American Institute of Steel Construction, AISC 360-16, Specification for Structural Steel Buildings; 2016

[15] American Concrete Institute, ACI 318-19, Building Code Requirements for Structural Concrete and Commentary; 2019

[16] Michael S. Troitsky, Planning and Design of Bridges. John Wiley & Sons; 1994

[17] American Association of State Highway and Transportation Officials, AASHTO, LRFD Bridge Design Specification; 2017

[18] Picture of concrete piers with reinforced steel corrosion problems. Available from: https://www.researchgate.net/figure/Concrete-structure-deteriorated-by-the-corrosion-of-steel-reinforcement_fig 1_265553963

[19] Picture of a bridge of structural steel with corrosion problems. Available from: https://www.researchgate.net/figure/Corrosion-of-steel-a-Extreme-corrosion-in-a-bridge-b-Railway-rolling-stock-being_fig 6_323826793

[20] Picture of a bridge concrete construction joint damage. Available from: https://www.researchgate.net/figure/Damage-at-a-joint-on-the-bridge-deck_fig 3_281864780

[21] Picture of a girder with fatigue fracture. Available from: https://www.researchgate.net/figure/Fatigue-fracture-failure-of-composite-beam-2-images-by-S-S-Badie-and-M-K-Tadros_fig 1_262973062

[22] Hernandez-Duque G, Schorr M, Carpio JJ, Martinez L. Preservation of the infrastructure in the Gulf of Mexico. Corrosion Reviews. 1995

[23] Picture of cathodic protection of reinforced concrete beam. Available from: https://www.vector-corrosion.com/blog/cathodic-protection-concrete-corrosion-prevention

[24] Budynas RG, Nisbett JK. Shigley's Mechanical Engineering Design. 10th ed. McGraw Hill; 2014

[25] Katthy Riggs Larsen, New Legislation Focuses on Extending the Life of Highway Bridges, Federal Highway Preservation (FHWA); 2008